WHY BIRDS SING

ALSO BY DAVID ROTHENBERG

Is It Painful to Think?

Hand's End

Sudden Music

Blue Cliff Record

Always the Mountains

WHY BIRDS SING

One man's quest to solve an everyday mystery

David Rothenberg

ALLEN LANE
an imprint of
PENGUIN BOOKS

ALLEN LANE

Published by the Penguin Group
Penguin Books Ltd, 80 Strand, London WC2R 0RL, England
Penguin Group (USA) Inc., 375 Hudson Street, New York, New York 10014, USA
Penguin Group (Canada), 90 Eglinton Avenue East, Suite 700, Toronto, Ontario, Canada M4P 2Y3
(a division of Pearson Penguin Canada Inc.)
Penguin Ireland, 25 St Stephen's Green, Dublin 2, Ireland (a division of Penguin Books Ltd)
Penguin Group (Australia), 250 Camberwell Road,
Camberwell, Victoria 3124, Australia (a division of Pearson Australia Group Pty Ltd)
Penguin Books India Pvt Ltd, 11 Community Centre,
Panchsheel Park, New Delhi – 110 017, India
Penguin Group (NZ), cnr Airborne and Rosedale Roads, Albany,
Auckland 1310, New Zealand (a division of Pearson New Zealand Ltd)
Penguin Books (South Africa) (Pty) Ltd, 24 Sturdee Avenue,
Rosebank 2196, South Africa

Penguin Books Ltd, Registered Offices: 80 Strand, London WC2R 0RL, England

www.penguin.com

First published in the United States of America by Basic Books 2005
First published in Great Britain by Allen Lane 2005
1

Printed in Great Britain by Clays Ltd, St Ives plc

A CIP catalogue record for this book is available from the British Library

978–0–713–99829–0
0–713–99829–6

For my father

The birds sing blissfully until the joy
is so great as to be unbearable:
this joy cannot be heard from afar,
but if you come near it, you will succumb.

— REB NACHMAN

Contents

Preface

IN 2000, the year I began to play music live with birds, I met a white-crested laughing thrush who changed my sense of how music emerges out of nature and how melodies can cross from one species to another. That encounter began the journey that is this book, an attempt to answer the beguiling question of why birds sing.

Some of you may consider it to be a simple question with an easy answer: birds sing for the sheer joy of it, because they can. Others may believe bird song to be fully explained by science: birds sing to attract mates, to prove their genetic fitness with exhausting displays of virtuosity, or defend their territory with angry sounds. Surely there exists some elegant, easy explanation for the beauty of bird song, perfectly consistent with the principles of natural selection? The mystery turns out to be far deeper. The humble question, Why do birds sing? forces us to reconsider what music is and where it came from, what sorts of thinking animals might do and to what extent we can communicate with them. Which of our human abilities are best suited to enter the minds of other species? We should not dismiss the possibility that pleasure in song may be something humans and birds can share.

The path toward an answer migrates through songs from birds far and wide, from single sneeps to half-hour arias, following the tiny marsh warbler from Belgium to Zambia and back, listening to the songs he picks up along the way. It leads through thick tomes of scientific data, including a two-hundred-page study of the wood pewee's song that contains only three notes. It calls for learning the difference between mockingbird and nightingale, neither to praise nor to mock them but to figure out how to honestly listen and join in. It takes us through poems that glean rhythm and meaning out of fleeting sounds never designed for humans to hear, through music from many human cultures full of inspiration from the avian world, and ends in an Australian rainforest, where I play along with one of the shyest yet grandest singers in the world, the Albert's lyrebird.

The National Audubon Society says that seventy million people in the United States—an astonishing one-quarter of the population—call themselves bird watchers. Many of us were first lured into the twittering world of birds through their sounds, but we often stop listening as soon as we identify what we cannot see darting through the trees. This book is for those who want to take a deeper listen. Slow down these sounds and you will hear much more nuance and structure. Print them out and they become complex rhythmic images. A small number of dedicated investigators from both science and art have scrutinized these avian songs in an effort to capture some of their spirit.

I come to this project from several sides. As a musician, I have always tried to develop my work through listening and learning from the natural world, while never quite sure what this really means in practice. I wish my music could be at once as spontaneous and as correct as what birds sing, although I don't believe it can easily be so. From imitating nature on my own instruments to playing along with prerecorded natural sound, I have moved toward direct interaction with birds as they sing.

As a philosopher, I have long been wrapped up in the question of what humanity must do to find a home in the natural world. The seemingly innocent topic of bird song shows us that we need a combination of many visions of nature to make sense of the whole. This book may present too many stories to offer a rigorous argument in philosophy, but it remains close to the field I first studied in the sense that I may love the questions more than the answers, which will never completely set the wondering to rest.

One of the questions that won't go away is this: Can any explanation for beauty be satisfactory? Despite all we have uncovered about how evolution is able to produce marvelous bird songs through generations of slow transformation, no knowledge tempers our joy. No theory makes the music go away. Nature is always wonderful, however much or little you know about it.

What fascinates me most about this question is how it illuminates the disparities among the many human ways of knowing. Information does not really touch experience. A lovely piece of music actually says nothing at all. Birds certainly sing to find love and to find home, but these reasonable purposes do not deny joy. If science is to comprehend happiness, then it should employ the skills of musicians and poets, who have used different human abilities to find meaning in the natural world.

In addition to the interspecies communication that happens when birds and people make music together, there is the further adventure of interdisciplinary cooperation. It is starting to happen. Here's an example from research on whale songs: Michel André and Cees Kamminga had been studying the rhythmic, clicklike sounds of sperm whales for several years off the shores of the Canary Islands when they made a recording they couldn't figure out. The scientists couldn't identify the individual whales simply by listening to all the clicks—too many overlapping rhythms! It's the same problem Western listeners have when they first hear a large ensemble of tight West African drummers. How can each player maintain their individual rhythms amid the great mix of patterns and beats? André and Kamminga hit on the idea of getting a Senegalese drum master, Arona N'Daye Rose, to help them.

With his highly trained ear, Rose could hear the individual statements inside the mélange, and was able to pick out specific whales' rhythms from the quick-beating fray. He determined that the profusion of clicks was organized around a single dominant beat. With the musician's help, the scientists concluded that each whale has its own distinctive click train rhythm, a result that no previous study had found.

If musicians can help whale song research, they ought to be able to help with birds. Complex bird songs—those of mockingbirds, nightingales, lyrebirds, and many others—share many structures with human music, from similar forms of organization to similar scales and phrasing. The history and reasons for such designs are better documented in music than in natural science. Yet science and music have generally believed that they walk different paths; one subjective, the other objective. A journey along both routes shows many opportunities where they might intersect, once they admit the value of each other's methods. This book introduces those few scientists who have tried to use music as a tool of analysis of the sounds of birds, but there is much more that could be done. I hope to inspire more scientists and musicians to work together in engaged interaction with the natural world.

I have written this book mostly in a stone cottage in a village surrounded by singing birds. At times I have listened so long that my ears seemed to play tricks on me. One day I was playing some recordings of the Eurasian curlew to some composer friends, comparing the swooping mad climax of its tune with what Olivier Messiaen did with it in his "Catalogue d'Oiseaux." Later that night, I was startled awake, sure I was hearing that foreign seabird's song coming from the tree next to my window. How could this be? It must have been a musical dream. I had been listening too hard.

The next day I noticed one of the neighborhood mockingbirds back in my tree, singing with that strange enthusiasm they sustain for only a few weeks in autumn—long after mating and territory defense are done. So intent and extreme, one tumbling phrase after the next. In the middle of his aria I heard it again, in the plain light of day. Those same curlew notes, whooping up to the burble, exploding to the peak. Did he pick up the song from my tape heard just the day before? Or maybe he's just inclined to follow the same turns of phrase? As you read this book, you'll learn that both explanations are plausible.

Read on, and you'll find out what we know and don't know about the range of possible things birds may sing. Although I have tried to capture the lilting flavor of bird song in the language of this text, it is best to hear it for yourself. The ears take in sound before we have words to explain it. Go out into the spring wilds and hear these exuberant singers in their own habitat. You may also hear many of the songs described in the text on the web site www.whybirdssing.com, and learn how to get the CD that accompanies this book, also called *Why Birds Sing* (Terra Nova Music TN-0501), which presents the latest of my musical collaborations with the natural world.

THE WHITE-CRESTED LAUGHING THRUSH

CHAPTER I

You Make My Heart Sing

It is March 2000 and I am in Pittsburgh to jam with the birds of the National Aviary, the finest public collection of caged birds in the United States. The plan is to arrive at dawn, to catch the wary singers in their early morning chorus, when the most sound is happening. The artist Michael Pestel is waiting for me at the aviary gates, far from the fancy parts of town. Pestel has been playing with the feathered residents of the place for years. The human staff like to let musicians in during the early hours before the public, mostly guided schoolchildren, bring their own noise and chatter.

At six in the morning the doors are still closed. All kinds of shrieks and whoops come from inside the walls. Through the screens we see darting movements of huge dark wings. Pestel, looking like a disheveled artist not used to getting up at this hour—the scraggly beard, the uncombed grayish hair—has his flute and various homemade stringed instruments. He also has a bit of the explorer in him, his long untucked shirt with many pockets full of hunters' bird calls.

I assemble my clarinets and saxophones out of their cases, along with a large plastic Norwegian overtone flute and some Bulgarian double whistles. A bit drowsy, but ready to hear what these birds have up their sleeves, we head for the marsh room, a vaulted expanse with an observation deck and water birds from all over the world.

Sunbitterns and egrets, spoonbills and teals. A green oropendola swoops over the water and gray Inca terns, with their impressive white whiskers, walk gently along the railing. Splashing and plunging, calling, swimming. I strain my ears to catch some pretty rocking bird beats. They sound familiar. The aviary is blaring Marvin Gaye at top volume to these birds at six o'clock in the morning. They are definitely squawkin' and squealin'.

"I cannot work in these conditions," mutters Pestel. "We've got to get these people to turn that racket down."

"Didn't you tell them we were coming?"

"No," he shakes his head. "Art always arrives without warning." As an art professor, he should know. At first he was a sculptor, but living in the same city as this incredible aviary had led him into the world of music. Motivated by the presence of these flying musicians, Pestel over the years has picked up flutes, recorders, bells, whistles, anything else that the birds might respond to. Understandably, he has developed a unique style of playing, somewhere between Eric Dolphy and the South American musician wren. The music is just one part of an artistic oeuvre that includes gallery installations with bird sounds, pebbles, and revolving wood structures, set up in exhibitions all over the world.

I'm still concerned. "You sure they'll let us do this?"

"No problem, man, I've come here many times before. These people know me. These birds know me."

The sprinklers are turned down. Marvin is turned down. I wonder if they will prefer our live music. Does a blue-crowned motmot or a violaceous euphonia really want to hear strange instrumental shrieks before breakfast? Weren't they doing fine with "What's Going On?"

Wittgenstein had the nerve to warn us that if a lion could talk, we would not understand him. Can you be so sure, Herr Ludwig? If a lion *roars,* we do understand him. If a cat purrs, we understand her. And if the voice of an animal is not heard as message but as art, interesting things start to happen: Nature is no longer an alien enigma, but instead something immediately beautiful, an exuberant opus with space for us to join in. Bird melodies have always been called *songs* for a reason. As long as we have been listening, people have presumed there is music coming out of those scissoring beaks.

WE SET UP ON THE WOODEN DECK next to the railing separating us from the artificial swamp below. We listen out over the water, instruments warmed up, recording equipment wired and set to go. A black crow observes us with interest, sitting on a branch at the side of the room. He cocks his head, eyes us knowingly.

Pestel plays a long, low sliding note, followed by a scratchy puff of air. Something strange swoops down next to my feet, shuffles its large black wings. Some kind of ungainly turkey! I see from a plaque on the railing that it's a gray-winged trumpeter, captured from the Amazon, its quizzical gaze mute.

"What are you looking at?" I glare. He steps cautiously toward the microphone cable, ready to gobble it with a determined lunge.

"Hey." I brush him off. "Stop dancing. Sing." After a particular clarinet pitch, he proves his name: *Baaaph baaph baph ba, Baaaaaaph, baaph baph ba*. Sounds more like a trombone played underwater. Did I come to find music in this?

Call the sounds of bird life music and there's a place for humanity within them. Call them language and it's a foreign tongue with no hope of being understood. To write out the song of the rose-breasted grosbeak, Aretas Saunders had to use free-ranging neumes as well as unpronounceable syllables. Here are rhythms, tones, melodies up and around—none of them too close to our systems of music or language. They sound out from an alien mind. But there are millions of bird watchers all over the world, and we need something to help us keep track of these many distinct sounds.

The philosopher Thomas Nagel says we'll never know what it's like to be a bat because we can never take on bat experience from within, but only imagine it or reconstruct it. Same with birds: Who can know what they feel as they sing, listen, and sing again?

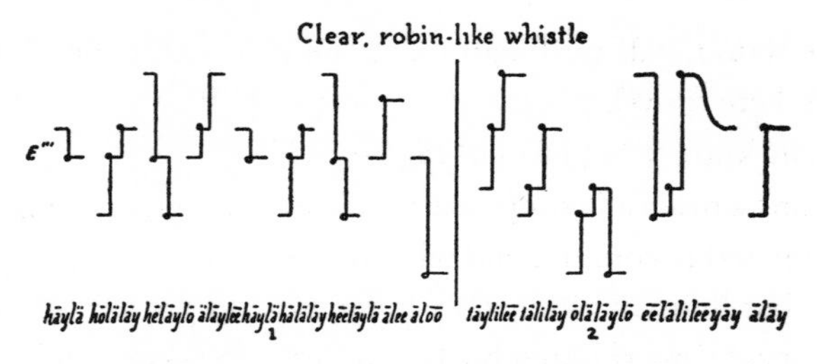

SONGS OF THE ROSE-BREASTED GROSBEAK

Yet who knows how anyone else responds to music—listening, playing, composing? The composer John Cage asked, "What could these three activities possibly have to do with one another?" Let's take bird song at face value, engaging with what it sounds like rather than obsessing over what it's supposed to mean. It don't mean a thing if it ain't got that . . . humanity? logic? story?

I don't claim to know much about music but I know what I like. I want to be surprised. I am easily bored and try not to play anything I have ever heard before. Yet how can I ever be so new? We are bounded by our memories and what we are born and have learned to do. Are we so far from birds, parroting expected phrases—songs either hardwired or learned by rote in youth in order to survive? Playing jazz for years, I learned a series of stock phrases and ways to turn around a scale from such masters as Parker, Coltrane, and my own teacher, Jimmy Giuffre, and now I'm supposed to mix and match this repertoire into the sudden game of improvisation. An outside listener—say, some superintelligent alien bird—might find human music merely a repetition and recombination of meaningless fragments and syllables; it has no message that can be broken down into a grammar of its parts.

All of a sudden there's a strange voice. A human voice? "Who," I hear. "Who. Who what where why. Who what where why."

It's that crow. Not just any crow. This one talks.

"Did you hear that?" I coax Pestel up from the flute, which he's been muttering into while blowing a tone, *drdrdrdrgdrdrgduh*—his signature sound, half bird, half man. "Oh," he said. "That's Mickey. He's been here for years."

"Does he know what he's talking about?"

A talking crow isn't supposed to know what he's saying. Parrots aren't supposed to comprehend what they imitate. They don't do it in the wild, but living with us, they know how to get noticed. Anyone who has spent time training birds will know they constantly surprise you, choosing precisely those sounds of yours to imitate that they know will catch your human attention.

But we're not here for this. We want music. Answers, not questions. Mickey can talk, but can he sing? *Caw,* he pipes up emphatically, somehow far too humanly: *Caaaaww*.

"We don't want that either." Pestel shakes his head. "This bird is just imitating people imitating crows."

A hyacinth macaw is paying attention to us. We're moving as we play; he's moving back and forth too. Swaying in time to the music. He commands attention on a perch in the center of the room. But still no song.

The greater flamingo is fed up. He's pink, forbidding, with that backward twisted neck. *Brahh Brahh Brahh Brumphphph,* he croaks. It's so loud it puts all the other marsh birds to shame. The cacophony rises. Is it a wild swamp scream symphony? Or just a vocal protest? Are we taking too much of their time?

"Man, that pink thing just won't shut up," growls Pestel. "I cannot work in these conditions. Let's move to the rainforest."

"But has the rain stopped?" I worry about the instruments.

"Don't worry, they'll turn it off for us."

IN THE RAINFOREST ROOM, we are engulfed in sweltering humidity. Mist lingers from the early morning rain. Instead of looking down to find birds, we look and listen above, all around us in the high canopy. The birds are smaller, sluggish at first, but as we start to play they dart all over. Loud, white Bali mynas, magical fairy bluebirds.

An austere Indian hornbill. The huge Victoria crown pigeon of New Guinea with its graceful stellated crest prances slowly on the ground. Blue-black superb starlings and green woodhoopoes. The birds are from all over the world, so the sound in this indoor Pennsylvania forest is a globalized mixture, nothing you would ever hear in the wild; it's a unique composition defined by confinement.

These tropical types are more agile, instantly melodic. *Ba ba bu ba pe pa,* sings a bright-yellow Taveta golden weaver in a pentatonic scale. Magnificently clear, five open tones. It's an open invitation to us wind players. All the world's human cultures welcome those five friends. We fumble, we test, we imitate. Does he care? He keeps singing his same sunny tune.

Soon he's eclipsed by the white-rumped shama, a virtuoso *explorateur*. One new phrase after another. Anything we play is just a challenge for him. An orange thrush of the tropics, this guy keeps coming back with a new variation. Whatever we feed him, he has a louder retort. Every song he sings seems brand new.

"Wait a minute, I thought these songs were innate," I ask Pestel. "Don't these guys need just one simple song sung as well as possible to do the job?"

"Calls," whispers Pestel. "Bird *calls* are innate. Those are the sounds they make with specific meanings: 'Where are you?' or 'I'm hungry' or 'watch out, a hawk's circling above.' Songs are something else again. If they're complicated, they have to be learned. And the birds can only learn these songs at certain sensitive times in their lives. Songs help them stand their ground and lure in mates, but they, like our music, don't have such a clear message."

"You mean they're born knowing sounds that mean, and have to learn the sounds that express?" It seems backward to me.

"You can explain it like that, but that story doesn't really touch the music." He pauses to blow some air through a nose flute. "Every bird's got a syrinx instead of a larynx. It's got two sides instead of one. They're not like us. They can sing several songs at once, and most have the potential to make far more sounds than they usually do.

Humans get involved, we provoke them. Witness the talking crow. Now stop talking and let's play."

With instruments poised at our lips, Pestel and I move slowly through this man-made forest, with those fake drips from the real leaves falling every day. Looking, listening for particular birds who are ready to interact with us, to take us seriously as singers in the dawn chorus.

In front of one thicket, I play a few notes, and all of a sudden a strong, rhythmic outburst comes out. *Brr du du du.* I play something like it back: *Br du du du.* And then as I weave a melody the bird joins in above me: *Be pu be pu be pu beep!* Who calls in there? Hmm . . . he's gray, black and white, robin-sized, hopping, dancing around like mad.

I keep playing, he's responding. At first he comes back at me with rising arpeggios, strong and tough. I play back. He cocks his head, leaps to join in. My notes change. His notes change. There seems to be some connection here. What is the message? If it's music, the message matters far less than the sound. Will we go somewhere together that we couldn't apart?

A woman walks by, pushing a huge mop, swabbing the place down. She looks up with a smile. "Are you getting it on with my man up there?"

"Yeah," I say. "Who is that?"

"That's a white-crested laughing thrush."

"Oh yeah?" I laugh, and the bird laughs some more. His laugh is a melody, a saxophone laugh, a Charlie Parker laugh.

"Is he getting up there with you?" She laughs.

On the hillsides of their native Southeast Asia, these laughing thrushes go around in noisy, cackling groups of one or two dozen. Their sound is generally considered a call, with specific social functions, rather than a purely melodic mate-attracting or rival-repelling song. Both males and females do it. Does this mean my bird is trying to tell me something specific, to get me either into his group or out of his world? He seems to live on his own, apart from any other members of his tribe. Perhaps he is lonely. Maybe the distinction between song and call is not so clear when a bird is confronted with alien music. This guy's sounds are definitely changing in relation to mine. Something is going on.

Pestel saunters up and takes stock of what he hears. "Wow, I've never heard that bird so excited before. You seem to have gotten through to him."

"Am I friend or foe?"

"Careful," Pestel warns me. "Don't succumb too quickly to that evolutionary model." He barks on a bird call strapped to his flute with a rubber band. "The real world is always more than they tell us."

Hear bird sound as music and there is always some mystique to enjoy. Hear the whole *world* as music and you'll find we live inside a plethora of beautiful sounds. How many other creatures out there are waiting for the chance to jam?

It's hard to write about or describe any kind of music, much less that made by a species so far from our own. *Holalay helaylo heelayla,* a tremulous trumpet in the trees. It's ringing loud and clear. What more can we know? When asked, "Why do birds sing?" most scientists would answer that birds let out their melodies to establish territories and to make themselves attractive to potential mates. As I dive into the history of this research, it turns out to be a much more subtle and intricate story. There is no clear reason why house sparrows get by with simple sneeps and pips, while the brown thrasher needs thousands of distinct song motifs. Without that amazing need to sing, the thrasher's just a plain brown spotted bird high up in the trees.

Evolution is not supposed to produce beauty for the sake of loveliness alone. Science needs data to back up every claim. A responsible scientist won't pretend to know why birds sing so well; will he disagree with the idea that birds burst into song out of pure joy?

SINCE THE RISE OF MODERNISM we have come to accept all kinds of blends and twists of organized sound as music. This is the most powerful legacy of the age of abstraction in all the fine arts—just about anything can be appreciated for its inherent aesthetic qualities, from a blank slate or a concrete wall to a whoosh of wind, feedback, or electronic noise. Although this might make it hard to discern what is art from what is not, it works wonders for our ability to love all the sounds nature allows us to hear. If we are serious about new possibilities of natural beauty, then even the noisiest squawks and shrieks are ready to be taken in as music. We are better prepared to listen in to bird music than ever before.

Bird music has been around for millions of years longer than any human compositions. This alone should fill us with awe and give a sense of rightness and presence to the tunes birds try out on us. We expand our attentiveness to the world by taking their vocalizations seriously. Just because science demonstrates that a song has a specific territorial or sexual purpose doesn't mean that birds aren't singing because they love to.

Although the sound works of birds have many of the same attributes as human music—repeating patterns, themes and variations, impressive virtuosic trills and ornaments, scales and inversions—they also offer radical inspiration to musicians. With incredibly compressed forms, artistic statements using washes of many frequencies at once, and complex transformations of sound that only a syrinx can produce, they are meaningful and joyful at once, and the surrounding sound world is richer because of them. Music may be one form of expression that vastly different kinds of life have in common. All over the world—from Babenzélé Pygmies to Beethoven—there is human music derived from avian sounds. Musicians reverentially aspire to the innocent and tireless beauty of those leafy songs.

So go back outside, walk into the forest or field, take in the first bird sounds you hear. Don't worry who it is that speaks—you don't need to know the musician's name to catch the drift of the music. First, listen as a bird might. You're interested in only the sound of your own species, perhaps, and others come across as mere noise. We can never know, we cannot get inside the bird. But imagine this bird's life, encountering another: one of our own comes into ear-reach.

Lover, friend, foe? Is he crowding our space, is he luring us in? Is he threatening, or just standing his ground? That's seeing song as something practical.

Or else, imagine a bird enthralled with sound itself. His songs are beautiful, complex, clearly more than what is necessary to get the message across. There must be exuberance, there may be joy. The bird is endowed as a virtuoso and loves to show off, explore, and cry out. What an artistic life! Music may be the only language songbirds need to know. Their brains may be small but think how great a part of them is devoted to music, to pleasure, to art. Their song is the essence of delight every time it bursts forth, necessary and complete.

There is something inherently predictable about bird song, as there is a definite range of possible sounds each bird is likely to make. Why then do they all sound so sudden and fresh? Nature is never boring, but content with itself, without need for human restlessness—a haven for calm and contrast, placid and wild, all bounded by an essential chorus, each tone in place.

Bird songs are a genuine challenge to the conceit that humanity is needed to find beauty in the natural world. Whatever processes of evolution have led to their flourishing, no rigorous natural logic explains why they are so multifarious and complex. With deft listening, we can abandon our prejudices to find new expanses of music beyond familiar constraints. Their music is essential, not arbitrary; playful but purposeful; repetitive, not boring. It possesses the necessity to which human art aspires.

The laughing thrush keeps laughing with the clarinet. It's a jazz of the underbrush, an improvisation with the avian world. One animal's song reaches out to another. When music starts to happen *between* humans and birds, you don't have to peel apart the categories of *man-made* and *natural*—the interaction appears and grows before we comprehend it. As in the hottest jam session, it doesn't matter who's from where or who's played with who, it's the *sound* that counts, the rapport. Who better to coax in than a songbird with an unstoppable will to sing?

Pestel and I play for hours with the exotic musicians of the aviary. But they never stop, and we start to get tired and hungry. "Should we just let the birds sing and take our own humble place at the edge of the trees?" I wonder. Or at least go get some breakfast.

"No, we have to stay with them. Can't you see they're daring us to keep up?" Pestel is sure.

I hope he's right.

WHAT A NARROW SENSE OF MUSIC it is that only lets people in! We may widen the realm of art, just as we expand ethics to include the environment, and honestly find a way for our species to care about the rest of this fragile world. No less austere a fellow than Immanuel Kant thought it prudent to remark upon bird song in his great manual of aesthetics, *The Critique of Judgment*. Why, wondered the great rationalist, do we never tire of listening to the simple melodies of birds, whereas if a human being were to take two or three notes and repeat them endlessly, we would soon get fed up with it?

Bird song, Kant decided, was not really beautiful, but *sublime,* something wonderfully alien to our understanding—beguiling, but always beyond our reach. He surmised there is something most powerful about the pull of nature's shapes and sounds: they are wild, irregular, bold, shocking, and able to take us somewhere far beyond our merely human arts. We can hardly improve on them.

So that's it. No wonder this whole experience is making me wild. Playing with birds, rather than merely thinking about birds, I begin to feel what it is like to be a bird. I do not look for proof but only possibility, and hope for new ways to interact, new sounds to surprise. *Wild things.* The mind is never as powerful as the ability to sing and dance. *You make my . . .* The music happens before we say it's impossible. *You make everything . . .* The birds are listening, they too want more. Perhaps they sense something like, "Those people, they don't just cage us and feed us and listen to us—maybe they're ready to learn from us too."

Will science convince me that the music's all in my head, and not in the birds'? And what of poetry, which has long sought meaning by turning language into rhythm? And the long history of bird song making its way into the rules and passions of human music? Must I place bird song in a human context to draw any logic from it?

In an artificial rainforest in the City of Steel I met one white-crested laughing thrush who showed me how melodies can travel the line from one species to another. It is time to track down all the experts, to listen to what they have heard in the laboratory, in the wilds, in memory and myth. This book moves through these experts' knowledge and goes also to its limits, for no one answer is full enough to solve the mystery of why birds sing.

CHAPTER 2

To Drink the Sound

No one who listens would call a bird's sound random. There is pitch, rhythm, pattern, repose. These are structures you could find in human language, yet bird song is not so much like a language because there is no syntax. The sound summed together is the only message. Like music. What's music? Just defined: organized sound—for its own sake.

The forces and forms of the natural world ground all human dreams. We are born, we learn, we sing, we love, we mate, we raise children, we teach them, we grow old doing all these things, and one day die. Birds learn and love too, and it's all unfolded in their songs. Some need only one song, others have thousands. There is no easy reason why nature has both possibilities. And it is human nature to both wish to explain what birds do and to just sit back and listen in awe. Neither approach will satisfy us completely, as we too are restless, constantly flitting back and forth with what our minds can do.

When we turn bird sounds into words so as to better identify them, language gets twisted into rhythm. Start with simple attempts to turn bird sounds into words that might help us remember them. Here's an excerpt from a list of bird sound mnemonics, in alphabetical order by sound:

chup-chup-zeeee!	spotted towhee
chureee!	semipalmated plover
churrr, churrr (throaty, deeply trilled)	red-bellied woodpecker
churrrk (burry)	semipalmated sandpiper
chu-wee, chu-wee	yellow-bellied flycatcher
chu-whee, cheer-ee-oh (thin)	solitary vireo
chwee, chwee, chwee	American pipit
ch-wut	upland sandpiper
chyoo-chyoo-chyoo-tseee (last syllable burry)	cerulean warbler
chyup	varied thrush
click click . . . (typewriterlike)	yellow rail
come here . . . Jimmy . . . quickly . . .	solitary vireo
conk-a-reeeeeeeee	red-winged blackbird

Come here, Jimmy? Not really. The vireo isn't talking. We just want his incessant message to be something familiar. *Burry?* Means something like *husky*. To put it in our context, so as not to forget.

Many of the earliest and most accurate references to bird song appear in poetry. The poet in all of us is the part that gets most carried away when we listen to birds. Forget what this beauty might mean, and focus on how it makes us feel. We hear passion, love, exuberance, also melancholy and longing. Bird song is difficult to explain and far from our musical conventions, but ripe as inspiration for poetic wonder through many human generations. The poet might start there and move quickly from mnemonic to new rhythms and forms. In fact, the most accurate transcription of a bird song from the whole of the nineteenth century was made not by a scientist, but by a poet, the close observer of nature, John Clare. Poetic and scientific speculation on birds have been intertwined for two thousand years.

Among animals to watch, birds are the easiest to see, quickest to move, most interesting to follow. Your eyes and ears are all you need to track and to see them, which is why there is more nature writing on birds than on any other animals. The earliest human records of attention to the natural world show that in the past, people did not have a disinterested sense of figuring the world out for its own sake. They were more concerned with what the birds might mean *for us*. Of course they were important as food. But before the feast is their incredible music. In his great poem "On the Nature of Things" the Roman writer Lucretius had this to say about how people first heard them.

> Through all the woods they heard the charming noise
> Of chirping birds, and tried to frame their voice
> And imitate. Thus birds instructed man,
> And taught him songs before his art began.

So the sounds of birds, with their incessant rhythms, dances, and energy, were heard as the ancestors of human music. Our histories and varieties emerge, but for birds the song remains the same once a species' identity has evolved, eons before we pause to notice it. Imagine two thousand years ago hearing the fresh song of the nightingale, knowing at once it is a nightingale, and then imagine this same song hurtling toward our ears even further back to the roots of human time, sensing that it was there at the dawn of our species, at the moment we first decided to make our own music, to define our own song.

It is easy to think of bird song as a kind of "primitive" music and leave it at that. But it somehow outlasts all our own attempts at getting the tune right. Our music comes and goes with the warp and woof of culture. Humanity will never "get it right," because we are the species that is always revising how we live and what will work for us. What songs of ours will be remembered a thousand years hence? Very few. But the white-throated sparrow will still *Old Sam Peabody, Peabody, Peabody* his way through life. The eastern towhee still sings *Drink your tea!* The sounds of nature deserve respect because they are right. They work. They will keep working—unless we carelessly destroy them.

For a long time people have realized birds sing even after mated and sated, chicks flown from the nest, music still heard on into the autumn months. They sing because they must. Because it makes them what they are. Does the singer keep wanting something that will never be attained? Is he only the music while the music lasts? Human life is not like that of birds. We are never as sure as they.

This is music of birds, not for people. It seems exciting and relevant, but you can't get through to it. It is not meant for you. In "The Singing," contemporary poet Kim Addonizio hears a bird whose singing never stops, and wants it to be sung for her.

> . . . I could say it's the bird of my loneliness
> asking, as usual, for love, for more anyway than I have,
> I could as easily call it
> grief, ambition, knot of self that won't untangle, fear of my own
> heart. All
> I can do is listen to the way it keeps on, as if it's enough
> just to launch a voice
> against stillness, even a voice that says so little,
> that no one is likely to answer
> without anything but sorrow, and their own confusion.

Once it was enough to simply praise the beauty of a bird's sound, to be happy they are there. Now, tentative in our own longing, we wish for something real from nature's noise. Can these songs mean *anything* for us if they are sung bird by bird, one to another, while we are only casual eavesdroppers choosing to tune in for a lark?

Cultures that live on closer and more visceral terms with birds hear different things. "You may call them birds," said the Kaluli people to anthropologist Steven Feld when he began fieldwork among them in New Guinea in the 1970s, "but to us, they are voices in the forest." The sounds of birds offer many specific messages to the Kaluli. Some announce when certain crops are ready, something quite practical. Others are just talking, from the chanting scrubwren who says *sei yabe* ("a witch is coming") to the brown oriole's *wefio ḳum,* something like "shut up, retard." Did the Kaluli learn these words from the birds? "Oh no," they told Feld. "The birds copied us."

Very few birds ever learn to sing. Among the thousands of birds on Earth, most are born with the sounds they use. Only four out of the twenty-three major groups of birds learn to make their sounds: songbirds, parrots, hummingbirds, and lyrebirds. Most are closed-end learners who lose the ability to learn new songs once they reach adulthood. A smaller number of birds are open-ended learners, such as mockingbirds, starlings, and canaries. So we have reason to doubt the Kaluli's view. But Feld points out there may be more to the story. This is a culture whose whole aesthetic sense is wrapped up with birds. If a person sings a most beautiful song, those listening say, "He has truly become a bird."

Other peoples find different aspects of bird songs the most enthralling. The Temiar in Malaysia are more aesthetically impressed with thrumming, repetitive booming sounds that we would not consider particularly musical. Pulsing sounds in the dense jungle are incessant and invisible. These tones and repeating rhythms lure human listeners into a state of trance. They make the Temiar long for the world of spirits.

The Hopi recognize deep information in bird song. It was the mockingbird who brought all the languages to different human tribes, saying, "You shall be a Navajo, you shall be a Hopi, you shall be a Pueblo, you shall be a White Man." The bird sang songs still chanted at Hopi ceremonies today. The Chukchee of Siberia speak of the "attainable border of the birds," the place where they cross from summer to winter climes, a barrier in the sky, a dangerous crossing, but something actually accessible to the striver, the seeker. We too want to go there. The flight of the birds call us on. Their wings and songs show us the way. "Birds," said ecologist Paul Shepard, "are not *like* ideas. They *are* ideas." Anything we say about them is a human abstraction from the truth: Neither science nor art can convey the actuality of birds' experience from within.

The first famous English poem uses bird song to praise the arrival of spring: "Sumer is icumen in, Lhude sing cuccu!" Almost alien, almost near: Springtime is a-comin' in, Loudly sing *cuckoo*! By Elizabethan times the nightingale had risen from admired musician to deep symbol for endless, untiring love. "O 'tis the ravish'd nightingale," wrote Lyle in *Alexander and Campaspe* (1584), "Jug, jug, jug, jug, tereu! she cries." And Nashe in *Summer's Last Will and Testament* (1592): "Cold doth not sting, the pretty birds do sing / Cuckoo, jug-jug, pu-we, to-witta-woo!" Andrew Marvell (1681) loses himself in "The Garden" as he wafts into the trees:

> Casting the body's vest aside,
> My soul into the boughs does glide;
> There like a bird it sits and sings,
> Then whets, and combs its silver wings

He hears the song and wants to soar out of there, be lifted up into the winged world. Birds are the quickly moving visible parts of nature, and we want them to be speaking to us, to have a message, to lure us out into the real world.

In 1580 the great essayist Montaigne wrote of how birds learn their songs, and what he said is fully compatible with modern science:

> Aristotle maintained that nightingales teach their sons to sing, spending plenty of time and care at it. . . . We can deduce that singing is improved with training and discipline. Even among wild birds, songs are not equal to each other; all birds learn according to their own skills. During the period of learning birds may be in competition with each other. . . . The younger ones meditate and start to copy certain strophes; the pupil listens to the maestro's lesson and then repeats it with care, knowing just when to stay silent.

Today science dissects the pieces of the avian brain and discovers exactly which parts fire up to make the song. Amazing advances have been made since Montaigne's basically correct observations, but we still don't know why the birds choose the particular intricate structures that they do. We can only confirm that they work, that their listeners respond just so, that one structure is more probable than any other. Beauty is not easily explained away. Beware of pulling at angel's wings.

In 1690 John Locke voiced the great quandary of bird song that is still with us: If nature is so regimented and efficient, are not the birds "wasting their time" by constantly singing, exploring, and in many cases varying their song? "These birds are able to sing just for singing's sake, expending the same energy as if it were a matter of life or death." Perhaps it *is* a matter of life or death. Evolution wants to teach us that music is the fuel of a bird's life, but it has a hard time explaining why there is so much more music in it than apparently there needs to be.

Scientific attentiveness to bird song began in the Enlightenment, perhaps with the breakthrough transcription of that Jesuit polymath Athanasius Kircher, who had his hand in just about everything, from medicine to science to the deciphering of ancient manuscripts. No surprise that he wanted to decipher the sounds of the animal world. Kircher was among the first to try to transcribe a complex bird song in musical detail, in this case the imitative excellence of the nightingale. *Luscinia,* he wrote in 1650, is able to mimic with astounding excellence the sounds of the cuckoo, swallow, hoopoe, tawny owl, and quail—in fact, nearly all the birds in his surroundings "in an orderly manner," to the delight of those humans who happen to listen. But "their voices are not destined to entertain human beings, they simply express the emotions of *their* soul."

The eighteenth century also saw the greatest interest in people playing music along with birds. Keeping caged birds was a national fad in England and Germany, and special flutes were designed to enable bird keepers to teach their birds a fine repertoire of pleasant tunes. The most popular instrument to play to birds at the time was the recorder, also called the flageolet or block flute, which is much easier to play than transverse flutes, on which the player must blow across an open hole rather than straight into the fipple. One of the early meanings of the verb *to record* was "to learn a tune." Ornitholo-

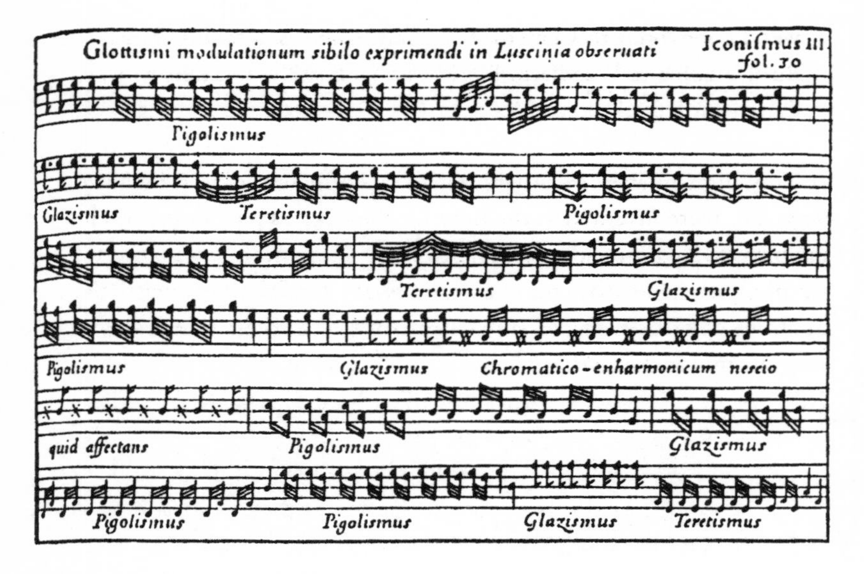

KIRCHER'S 1650 TRANSCRIPTION OF A NIGHTINGALE SONG

gist Daines Barrington, in the late eighteenth century, used the word to explain how birds learned to sing. "The first sound is called *chirp,* the next is a *call;* the third sound is called *recording,* which a young bird continues to do for ten or eleven months til able to execute every part of his song. When perfect he is said to sing his song *round.*"

A few decades earlier came the famous collection of recorder tunes called *The Bird Fancyer's Delight,* first published by Richard Meares in London in 1717. These songs were specifically designed to teach birds to sing. Only one copy of the original remains, in the Library of Congress, but in the 1950s it was reprinted and has become standard repertoire for several generations of new recorder players. The title page of the book explains colorfully what it is for:

> Choice Observations and Directions Concerning ye Teaching of all Sorts of Singing-birds, after ye Flagelet and Flute, if rightly made as to Size & tone, with a Method of fixing ye wett Air, in a Spung or Cotton,

with Lessons properly Composed, within ye Compass & faculty of each Bird, Viz. for ye Woodlark, Blackbird, Throustill, House-sparrow, Canary-bird, Black-thorn-Linnet, Garden-Bull-finch, and Starling.

Tiny high-pitched recorders less than six inches long, called *bird flageolets,* were invented at the time for the express purpose of teaching birds to sing these particular melodies. The tunes by and large are simple, jolly, and diverse, with only slight tendencies toward the qualities of birds who are supposed to warm to them. Here is one song meant for the starling:

SONG TO BE PLAYED TO STARLINGS,
from *The Bird Fancyer's Delight* (1717)

As we will see in Chapter 5, real starlings sing far stranger sounds than this. Could birds really learn such harmonious, unbirdlike tunes? In Germany, academies to teach bullfinches lasted for centuries up until a few decades ago. The pupils were usually kept in "classes" of about six birds apiece, caged in a dark room, "where food and music are administered at the same time." As soon as they started to copy a few notes of the song played in the dark, a little light was let in. Some teachers chose instead to starve their birds and keep them cloaked until they were able to sing what they heard, using torture to achieve beauty. This approach went on for at least nine months, during which time the birds really got their tunes down. They were brought to competitions and the finest singers were lavished with prizes.

German biologist Jürgen Nicolai has done thorough work in recent years on how captive bullfinches learn their songs. He was amazed to find that the birds seem to have an innate sense of how a tune should go, and tend to perform the songs better than their human trainers. If the trainer whistled unevenly, the bird evened out the notes. If the trainer started a tune and stopped it, the bird could finish the song flawlessly. All this for a bird who sings little more than soft squeaks and scrapes in the wild.

What did the music world think of the instruction of birds? Carl Philipp Emanuel Bach, Johann Sebastian's most famous son, wrote the following advice in an essay on the art of the keyboard: "Play from the soul, not like a trained bird!"

In the seventeenth and eighteenth centuries it was also fashionable to contemplate the origins of music. An entire scholarly field known as "speculative music" developed, and most imagined, on the evidence of Kircher and others, that human music emerged out of nature. Man looked and listened around him. In his *General History of the Science and Practice of Music* (1776), John Hawkins wrote that "the voices of animals, the whistling of the winds, the fall of waters . . . not to mention the melody of birds, all contain in them the rudiments of harmony, and may easily be supposed to have furnished the minds of intelligent creatures with such ideas of sound." Proof? The cuckoo not only announces the arrival of spring but sings a descending minor third. Hawkins believed the European blackbird to sing a fanfare exclusively in F major, though few listeners today would agree.

During this time the proto-Darwinian idea was in vogue that all human arts descended or appeared from nature. Yet in the years to come, our faith in our own art became less secure and we became more insistent that culture is something that distinguishes us from the rest of life. Music was later said to arise from something uniquely human—language perhaps, as a by-product of more practical kinds of communication.

I've already mentioned Immanuel Kant's famous *Critique of Judgment*. This is his third critique, after pure then practical reason, in which he outlines the characteristics of moral and aesthetic judgment. In this 1790 book Kant tries to explain why people find nature to be beautiful in an untouchable sense. He applauds the magnificence of the *sublime* in nature, from spewing volcanoes and awesome cliffs to thunderstorms and tremendous waterfalls. But, he eventually decides, the *beautiful* is better: pure, organized symmetry and art, a wholly human thing. We love sublime nature because it is so different from us. Beware—size and splendor too easily impress. Don't take them as seriously as form and perfection.

Kant never found the music of birds monotonous.

> Even a bird's song, which we can reduce to no musical rule, seems to have more freedom in it . . . than the human voice singing in accordance with all the rules that the art of music prescribes. . . . If exactly imitated by man (as has sometimes been done with the notes of the nightingale) it would strike our ear as *wholly destitute of taste.*

Tasteless, eh? We'd make it insufferable, but the bird in the bush is always perfect. Later Kant imagines children hiding in the woods, pretending to mouth the tunes of birds. The minute we find out the sound is not real bird song, we no longer appreciate it. "It must be nature, or be mistaken by us for nature, to enable us to take an immediate interest in the beautiful as such." This seems ambiguous: Do we love the bird's song because we love the song or because we love the fact that a bird sings it?

Such conjecture had already faded from history by the beginning of the nineteenth century. As science sought greater discipline, it had much less to say about bird song since it had no tools worthy of its rigor until the twentieth century brought accurate sound recording technology. Romantic poets took up bird music not as a clue to our deepest origins but as a symbol of inner feeling. They became the era's most prominent guides to our wonder, intoxication, and frustration with the revelation of beautiful bird sound and the emotions it might lead us to.

John Keats's "Ode to a Nightingale" of 1823 turns Kant's revelation of the untouchability of a bird's song on its head. Keats gets mostly depressed listening to so essential a sound, one that has resounded from the bush much longer than any lonely human cry for love or meaning:

> Thou was not born for death, immortal Bird!
> No hungry generations tread thee down;
> The voice I hear this passing night was heard
> In ancient days by emperor and clown. . . .
> Adieu! Adieu! thy plaintive anthem fades
> Past the near meadows, over the still stream,
> Up the hill-side; and now 'tis buried deep
> In the next valley-glades:
> Was it a vision, or a waking dream?
> Fled is that music: —Do I wake or sleep?

The poet is bothered that the paean of the nightingale is not for us. We are nothing next to your song, so what is the point of going on? In melancholy he soon forgets the bird and turns within, to his own problems.

Samuel Coleridge shied away from such a despondent view of this most unstoppable singer. His nightingale poem is several decades older than Keats's, but already he is suspicious of what too many heard as pining. "A melancholy bird? Oh! idle thought! / In Nature there is nothing melancholy." No, these night melodies are bursting with love and competition:

> 'Tis the merry Nightingale
> That crowds, and hurries, and precipitates
> With fast thick warble his delicious notes,
> As he were fearful that an April night
> Would be too short for him to utter forth
> His love-chant, and disburthen his full soul
> Of all its music!

Coleridge tells us more than Keats about the night bird's actual ways. *He* is singing, never she, and the males goad one another on in bouts of matching, contrast, and decision, presumably vying for the female's rare attentions, though that will be a matter for science to sound out a century hence. Love! love! love! *Jug jug jug,* the rhythms are rough-hewn, not sweet.

So far the depth and verve of this night music's clicking, searing rhythms has not been caught by either poet. One has to look to a third great romantic, John Clare, who heard the nightingale's rhythmic coos as nature's central poetic sound, the sweet beat that set his whole sense of rhyme aloft. In May 1832 Clare noted down the sounds of a nightingale singing outside his window in an apple tree. This turned out to be the most accurate rendering in words of any bird's voice for nearly a century:

Chee chew chee chew chee
chew—cheer cheer cheer
chew chew chew chee
—up cheer up cheer up
tweet tweet tweet jug jug jug

After writing down this bird's rhythms and tones, he worked them into several of his best poems, including "The Nightingale's Nest" and his masterpiece of self-definition, "The Progress of Rhyme." In this manifesto for his art, Clare describes how words and rhythm come to him from his daily work in the fields and his attention to the brilliant sounds surrounding him. The most salient sound is the nightingale's voice:

The more I listened and the more
Each note seemed sweeter than before,
And aye so different was the strain
She'd scarce repeat the note again:
"Chew-chew chew-chew" and higher still,
"Cheer-cheer cheer-cheer" more loud and shrill,
"Cheer-up cheer-up cheer-up"—and dropped
Low "Tweet tweet jug jug jug"—and stopped

One moment just to drink the sound
Her music made, and then a round
Of stranger witching notes was heard
As if it was a stranger bird:
"Wew-wew wew-wew chur-chur chur-chur
Woo-it woo-it" could this be her?
"Tee-rew tee-rew tee-rew tee-rew
Chew-rit chew-rit"—and ever new— . . . .
Words were not left to hum the spell.
Could they be birds that sung so well?
I thought, and maybe more than I,
That music's self had left the sky
To cheer me with its magic strain,
And then I hummed the words again
Till fancy pictured standing by
My heart's companion, poesy.

This poem may be the best example of how turning bird sounds into words—even what seem to be nonsense words—can make more sense as poetry than music. Kircher's earlier transcription is suggestive but it doesn't really read as music. Clare, with his faithful words, finds a true way to drink the sounds. It is clear from this poem that he really listened to the birds as he worked the fields, seeking meaning in the rhythms of the real world. Sadly, he spent his final years in and out of the madhouse, perhaps unable to reconcile what he was able to hear and what he himself needed in order to live.

Poetry could make sense of bird song meter at a time when science and music could not. The nightingale's repetitions and variations suggest tones and rhymes: *whit, bit, cheer, fear, new, you, few, still, shrill.* The song goes on for hours, it is the same and not the same, the endless machinations of the singing mind, a music of noises that means nothing more than the tumble of sounds. Clare's poem takes the free beat of the nightingale's song and bends it into regular eight-beat tetrameter. It is formal, but not simple.

These English romantics were flummoxed then floored by the beauty of these alien bird strophes and meters, which threw them into despair, awe, and raw inspiration. As honorable and melodious as the nightingale's song comes out in poetry, the song itself is still rather strange and crackly, musical in only a foreign way. That is what makes it infinitely interesting. I will return to this bird in Chapter 6, when more advanced tools allow us to listen more thoroughly—but perhaps more abstractly—to a sound that meant so much to an important generation of poets and dreamers.

In America the closest sound to the tireless verve of the nightingale is the song of the northern mockingbird, *Mimus polyglottos,* the vigorous mimic of many tongues. The mockingbird assembles its imitations of other birds' tunes in a specifically mockingbird manner, with easily identifiable repetitions of fours, fives, sixes, and sevens, often easily spaced, mixing repetition and contrast in really musical ways. You would not easily confuse a mockingbird's song with anyone else's once you know it, because of the uniquely ordered way in which his riffs are put together.

Once again it was a poet who made the most out of this bird's exuberance—that most American of bards, Walt Whitman, ever singing himself the joy of being alive. No wonder he was the first to really latch on to the mockingbird's rocking, to find beauty in it, not the insult or jeer that the bird's name inspires. His most famous poem rolls straight out of the mocker's own calculated tumbles and breaks:

> Out of the cradle endlessly rocking,
> Out of the mocking-bird's throat, the musical shuttle,

The three, the four, the six, Whitman knew it, knew it was the male who sang to the female, but the little boy narrating the poem wants the song to be for him alone, singing *for me, for me, for me:*

> Shine! Shine! Shine!
> Pour down your strength, great sun!

While we bask, we two together. . . .

Blow! Blow! Blow!
Blow up sea-winds along Paumanok's shore;
I wait and I wait till you blow my mate to me. . . .

Soothe! Soothe! Soothe! . . .
But my love soothes not me, not me.

Loud! Loud! Loud!
Loud I call to you, my love!

Picture Whitman, out in the breeze, taking all of experience in, wondering how to express its greatness. He sought new rhythms, new forms, new shapes, listening and looking everywhere he went. That quintessential American poet found the real rhythms of his continent and his time. The New World sings forth its colors as image of the coming century of tumult and change. The mockingbird is not the nightingale, subject of centuries of delicate Old World hymns to its magick, sweet midnight flutes and dark explorations. Out on the branch the mockingbird beams forth so that all can see and hear him, this gushing of energy and choice. What beats, what surety, what earthly power! If you want a new form for your poem, you need only listen hard and long, breathe it in, and try not to mock its rhythm but hold on to the way it grabs you and doesn't stop.

With this song the possibility for love arrives in the boy's heart, and in cascading, relentless mockingbird-inspired rhythms the poem tells the musical story in exclamations of joy, not really in anything so straightforward as words. He's already put them in italics because they are not quite words but stops, points, sounds:

O throat! O trembling throat!
Sound clearer through the atmosphere!
Pierce the woods, the earth,
Somewhere listening to catch you must be the one I want . . .

O past! O happy life! O songs of joy!
In the air, in the woods, over fields,
Loved! loved! loved! loved! loved!
But my mate no more, no more with me!
We two together no more.

Hearing the astonishing song, the peal of rhythms and twists of other birds' tunes that rhythmically rise up into the sky, the boy tries words but is left with inner melodies that speak of love. This greatest wonder of life begins to surge inside him, those "thousand warbling echoes started to life within me, never to die."

Listening out my own window to a singing *Mimus* on a budding bush, I realize how much this poem owes to the foreign music of the mockingbird. Here is a poet who heard resounding human stanzas formed upon avian rhythms, stretching the rules of poetry in his time to their expressive limits. He is one man who gamely reached for the attainable border of the birds.

Mockingbirds and nightingales have sung as they do for millions more years than human beings have been around to wonder why they do it so tirelesssly. Singing because of the need to sing. Listening for the prospect of listening. There is so much more inside these natural melodies than any reason could contain. Poetry finds verbal energy in the tumbling patterns birds have recited for centuries, but it does not touch one essential enigma: Where did all these beautiful rhythms come from?

THE LYREBIRD

CHAPTER 3

She Likes It

THE ALBERT'S LYREBIRD OF AUSTRALIA is one of the most impressive of all birds, with long curved plumes like the ancient Greek lyre and a powerful song to match. He's one of few to combine a magnificent appearance with an awesome and exact courtship display. Beneath his bouquet of a tail he's like a small brown pheasant, running quietly through the Queensland mountain rainforests, silent and inconspicuous, until he decides he wants to be heard, which is every day during his winter breeding season. Then he performs one of the most precisely choreographed rituals in the whole world of birds.

Every morning, just before dawn, he descends from his resting perch high in the trees and searches through the forest for a very special kind of display site, one of five or six he has identified in his territory. Unlike his more common cousin, the superb lyrebird, who performs on one-meter-round open mounds that take several weeks to construct, the Albert's lyrebird finds his stages ready-made, like Marcel Duchamp. He needs a spot where a mass of thick vines hang close to the ground then swing back up into the trees. Perched on one of these perfect spots, he begins his show.

He tosses his shimmering lyre-tail feathers tight over his head like an unbrella, so you can hardly see his face. Like a matador disguised behind a cape, he starts with a territorial call, announcing his place:

Breep, booua, bwe, ba boo pu tee! Then he begins a series of flawless imitations of many of the other birds that share his home—satin bowerbirds, rosella parrots, yellow honeyeaters, kookaburras. Although some birds learn to sing in a matter of weeks, it takes an Albert's lyrebird at least six years to successfully learn this song. They live for up to thirty years. In any forest group of thirty or forty birds, they all end up with the same series of mimicked sounds in roughly the same order, after those many years of practice that signal the arrival of maturity.

After several cycles repeating this song of imitations he embarks on a back and forth kind of dance, with his two feet on the vines, shaking them enough so that the trees high up behind and in front of him also shake, and he emits a different precise rhythmic music, original and exact, *gronk gronk gronk brr brr brr brr brr.* The whole forest shudders with his dancing. His head is invisible, the feathers embrace him. And oh—did I mention the bright red tail feather that he provocatively presents from behind? It sticks straight up in quivering invitation. When he is done, he starts again. After a few rounds he'll take a break to claw the forest floor with his large talons, looking for roots and grubs to eat. Then he'll move on to the next platform to start the whole performance again.

In the world of natural selection, a song and dance as spectacular as this can only mean one thing—this display is what female lyrebirds like to see and hear. Generations of female preference have led to the survival of appearances, songs, and behavior that might otherwise seem excessive and extreme. Why so magnificent a dance and so elaborate a song? Because of the elusive female lyrebird. She likes it.

But the chances that the lyrebird's impressive show will lead to mating success are quite slim. Female Albert's lyrebirds lay a single egg only once every two years. There are not so many of them around, and competition for their attention is fierce. The male sings, dances, and performs every single day, all day, of its winter life. In the summer his gaudy, swirling tail feathers fall off and he wanders the forest placidly eating, not troubled by the push of hormones. Next winter the performance is up on the stage of vines again.

Why does one single species of bird need to work so hard to succeed in life? Such extreme stories from the world of birds gave Charles Darwin some trouble. His theory of natural selection is one of the triumphs of science because it employs a very elegant and simple mechanism to explain the evolution of life's many ways, including elaborate behaviors such as the courtship of lyrebirds. But it doesn't quite explain why such excessive, seemingly inefficient behavior is necessary. Even though natural selection is one of the great conceptual achievements of modern times, it has a hard time explaining why birds sing.

The amazing diversity of life on Earth is one of the most wonderful things we can experience in the natural world, and it is easiest to admire in birds. Falcons are strong and swift, the better to nab a skittish rabbit on a field. Crossbills have beaks that close unevenly, the better to open up pinecones. Woodpeckers can get the most out of the barks of trees. These features seem easy to explain within the "survival of the fittest" idea that is at the heart of natural selection. Through countless generations, nature has selected for those traits that enable specific birds to succeed in life. Species develop their identities as certain qualities fit them to specific ecological niches. All life forms are interconnected in the way they prey on or support one another. When conditions change, species may become extinct. Others evolve to take their places.

Looking for efficiency, you can see much of nature as a paradise of precise engineering. The evolution of such fitness through generations of inadvertent testing, with no grand master in charge, is a wonderful discovery that well explains how complexity can have arisen without a guiding hand behind it. Darwin's idea offers scientific evidence linking humanity with the rest of the living world. Of course, not everyone is convinced that his cascade of randomness and emergent order makes any more sense than the traditional view, where biological diversity is evidence of a supreme being offering us the gift of a beautiful nature as proof of His existence—reason has never had much to do with faith. I'm not going to argue for or against anyone who says beautiful birds and beautiful songs are proof of God's works. But if you want to trust the evidence, let God exist through evolution—nature is all the more amazing the more we learn how it works.

Standing pat with God as the answer puts a *stop* to the question, Why do birds sing? It's too easy an answer for me. Evolution does not erase meaning and feeling from nature, except if we oppose the complacent "God made it that way" with an equally simplistic "natural selection made it that way." The world is not perfect. Nature seldom solves problems by the simplest means. Evolution, like God, works in mysterious ways. We must be wary of what our theories have a harder time explaining.

Why does the Albert's lyrebird always do its dance atop low-slung vines? Only to create a bigger ruckus as he shakes the treetops? Why do peacocks have such huge tails—the classical example of avian excess? Why do hornbills have horns on their beaks so large that they can hardly see? Why would a mockingbird sing for hours at a time when no one is around to hear? These are not obviously useful adaptations. If they are the result of years of selection, there must be something other than mere efficiency at work in nature.

What did Darwin say to this? He was convinced that females of many species have an innate aesthetic sense, and prefer certain traits simply because they like them. In *The Descent of Man* he wrote that birds "have strong affections, acute perception, and a taste for the beautiful." Not necessarily the same taste that humans have, but taste all the same. Bird songs "charm the female," and this is the basic idea our culture has inherited from Darwin's beliefs. Sexual selection is a subset of natural selection—it refers to those "arbitrary" characteristics that females prefer generation after generation, thereby reinforcing those traits in the population. Crazy acts, outlandish tails, unending, unlistened-to songs. Female choice for the sake of choice. Excessive? Ridiculous? Foolhardy? There is no accounting for taste.

Birds certainly sing while looking for mates, and people do as well. There's some truth in it, as a recent book in biomusicology called *The Origin of Music* also suggests. Jimi Hendrix, says psychologist Geoffrey Miller, was a great guitarist, and he had many girlfriends. "His music output did him no survival favors. But he did have sexual liaisons with hundreds of groupies. . . . Hendrix's genes for musical talent probably doubled their frequency in a single generation through the power of attracting opposite-sex admirers." A strangely eugenic view of a great musician. Jimi didn't get that good by doing what girls liked to hear. He spent more time alone experimenting with his guitar. He had to get into the music itself, like the lyrebird who works on his elaborate song for years. The beauty of the music that results is not acknowledged in the evolutionary explanation.

I am surprised to learn how much aesthetics mattered to Darwin. Beauty must be *loved* by nature to be found so often—perhaps he was a romantic at heart, bound by the sentiments of his time, as expressed in the poetry of his contemporaries Coleridge, Keats, and Clare. To a poet or artist it is comforting to hear that the greatest of all biologists with the most revolutionary theory should accept that beauty and taste are essential for evolution to proceed. Chalk one up for culture, in both animals and man.

At the same time, critics from within science, in Darwin's time and after, have been confused as to why the founder of evolutionary biology should have included such an enigmatic idea without wanting to develop it. So sexual selection is based on female choice, is it? How and why should *this* behavior have evolved? Many of the advances in our understanding of evolutionary biology have emerged from attempts to answer the question Darwin didn't want to ask: How can the beautiful be useful?

"That animals utter musical notes is familiar to every one, as we may daily hear in the singing of birds." Darwin knew there was an elegance to these sounds that no amount of deciphering of the song as language could touch. He distinguished bird songs from other bird sounds because they have that "more subtle and more specific effect which we call *musical* expression . . . , the delight given by its melody." Whereas a scream usually sounds like a scream, whatever animal is making it, a song is often more elaborate than it needs to be. Its beauty and identity lie in the details.

Darwin was sure that bird song served the function of sexual attraction and had this to say to critics who pointed out that birds often sing in the autumn, long after mating is done with: What for? Practice, practice, practice. "Nothing is more common than for animals to take pleasure in practicing whatever instinct they follow at other times for some *real* good." Male song is sung to be attractive to females, but also as a clear expression of a whole range of emotions, "such as distress, fear, anger, triumph, or *mere happiness*." No conflict between reason and emotion here.

Lest you think Darwin, with this blend of art and science in the minds of birds, to be unusually pancultural, note what he said to justify his belief that taste for the beautiful was by no means a hallmark of humanity—"Judging from the hideous music admired by most savages, it might be urged that their aesthetic faculty was not so highly developed as in . . . birds." Hah! What would he have thought of Stravinsky or Cecil Taylor? Too much musical commentary on bird song is just as trapped in prejudices of its era. In today's world of breakbeats and record scratches, many of those bird sounds previously heard as noisy or ratchety or "burry" might be more musical than they were to earlier ears. The song does not remain the same if we change the rules of what music can be.

So Darwin was no cultural relativist. That's no surprise considering his place and time. The masters of the British Empire still thought they could know and own everything. Darwin did recognize that what is beautiful to a pelican or hornbill might look ridiculous to us. He encouraged us to look and listen to nature and marvel at what works—not for any disinterested curiosity, but because he wants us to observe the way nature's males display to females. This should teach us something about human life and human origins. Two chapters in *The Descent of Man* are concerned exclusively with birds. We too are part of this inexplicable range of beauty.

Darwin says female birds have a natural aesthetic sense. How dare he! If nature is a domain of reason, then even beauty must have a definite purpose. Darwin's followers have all tried to wrestle with his acceptance of raw beauty in the world. A woodpecker's beak makes good sense, but what is the sense in the taste of a female mockingbird who likes a thirty-minute, effusive male mockingbird song?

In 1915 R. A. Fisher combined Mendel's genetics with Darwin's idea of sexual selection, suggesting that preferred traits can run away in an extreme direction because the females instinctively prefer good genes. If a female chooses a mate with fine feathers, or a fine song, she is choosing a trait that is largely inherited. So her sons will likely inherit this same trait. They will hopefully be the best singers around. Is it so clear what makes a song the best? It doesn't really matter, what's important is what the fashion is, and that species fashion does not fluctuate because the next generation of daughters will inherit their mothers' preference for whatever sounds and sights are in vogue. Outlandish plumage, extreme tails, wild and weird tones. Fashion may be fickle, but in animals it moves not in waves but in particular directions. These push forward and the extreme trait can be exaggerated from generation to generation. Fashion defines a species, and once a species is defined, it sticks. In extreme cases, you end up with eccentric features like those musical-looking male lyrebird tails.

Is it really useful to have such excess plumage? Don't the birds get stuck in the trees and make themselves easy targets? It just goes to show you what can evolve in habitats where there are few significant predators. That's why big islands like Australia and New Guinea have such bizarre birds—no one was around to gobble up lyrebirds while they evolved their intensive song and dance millions of years ago.

You might be a sensitive peahen and make the sensible choice of a peacock with a reasonable, short tail so he can move around quickly and efficiently through the forest. Fine, but your son will inherit the short tail and the next generation of daughters will inherit the majority's preference for long tails. Your son may survive longer but you are less likely to get any grandchildren, because he won't be so popular with the girls. Sexual selection is not simply adaptive natural selection. What is it then? Proof that if you care about your children you'll want them to run with the popular crowd. You don't want them to be too different or your genes will go nowhere.

How do such peculiar fashions create the key characteristics of a species? Fisher needed genetics to make it seem sensible, something unknown in Darwin's day fifty years earlier. The mother bird has genes for preferring a complex, specific song. Her children will inherit this preference gene *and* the song gene. The males will sing the "better" songs and females will prefer the "good" songs. Sexual selection links the preference for the trait with the trait itself, so both pass on to the next generation and every subsequent generation after that.

What makes one song better than another? According to Fisher's "runaway" theory, it doesn't matter. If the females start to like it, then it catches on. Some authors describe this process as analagous to pop music charts or best-seller lists, but I don't think such metaphors are all that accurate. These traits in nature reinforce themselves over a slow trajectory, while human culture dips and bends from one fad to the next.

Fisher believed that the majority's penchant for one kind of song can get started from a very slight preference in just a few birds. It might first emerge from chance fluctuations. It's positive feedback—success breeds success. It's best to go with the flow, do what the majority does. So extreme beauty in bird song and plumage does not result from what people might wish for: individual virtuosity and specialness. Instead, it is a matter of being what your species tells you to be. No need to try singing like a nightingale if you are a swamp sparrow. As a bird, it would never even occur to you. That's not the way of animal aesthetics.

So do female birds actually prefer mates that sing in a particular way? Since Fisher's original hypothesis, there have been hundreds of scientific experiments designed to test very specific cases of this and related ideas. We'll go into detail on a few of these experiments later, but for now let's be honest—the evidence on the whole is mixed. These few examples give a sense of the kind of preferences scientists have been able to test for: Swedish scientists placed recordings of collared flycatcher songs inside stuffed dummy birds strategically placed in nest boxes fitted with traps. Snap! Far more females were lured into traps with song tapes playing than those that were silent, suggesting they'd rather hear potential mates sing than sit mute. But what should they be singing?

In England it was determined that sedge warblers who sing many different kinds of songs mate earlier than those with a more limited repertoire. The females then prefer males that know a lot of tunes; they approve of diversity. Female canaries are more excited into building nests when they hear a large repertoire than a small one, and female song sparrows also prefer a large repertoire, but only in captivity—in the wild there is no correlation. In brown-headed cowbirds, females prefer the songs of dominant males over those of their subordinates.

Is this avian aesthetic arbitrary? Fisher thought that any odd thing females prefer might conceivably evolve into a defining characteristic for a bird species' song. His vision of nature leaves room for plenty of accidents. Just as science was uncomfortable with Darwin's acceptance of birds' innate connoisseurship of beauty, Fisher's followers would not rest with the idea that these fabulous ornaments in appearance and resonance should be the result of mere whim. Could the excess itself be some kind of evolutionary message?

In the 1970s Israeli biologist Amotz Zahavi formulated an idea known as the *handicap principle*. Why, during mating season, should male pelicans grow huge bumps on their beaks that make it nearly impossible for them to see the fish they are trying to gulp down as they plummet into the sea? The males, says Zahavi, are handicapping themselves on purpose. Not consciously, but by a design that is the result of sexual selection—they are sending a message of pride to their potential mates: See, I've got this huge bump on my beak, I can barely see what I'm doing, but I can *still* bring home the fish. I can take care of myself with this outlandish thing in front of my face, so I am strong enough to take care of you. Or at least mate with you. Or do whatever you expect me to do *and do it better*!

The handicap principle is extremely popular among biologists today because it provides *some* reason, however counterintuitive, for outlandish or inefficient behavior in a natural world that is still supposed to be following rigorous rules of cause and effect. It is a convenient way to explain why certain birds seem to spend so much time doing unnecessary things. Bowerbirds build immensely complicated artworks to serve as seduction lairs for choosy females. These structures take weeks to build and aren't even nests, just love shacks where mating will take place. Each species builds its bower in a characteristic shape, and the males endlessly tinker with their work to get it right. They look a bit like the earthworks of Andy Goldsworthy. The satin bowerbird must decorate his with objects colored blue: either flower petals, or, if he can't find those, blue plastic spoons that he may carry back from picnic tables miles away. According to the handicap principle, the peacock's ungainly tail or the lyrebird's way-out plumes are not fashion gone wild but genuine handicaps designed to demonstrate to females that males who possess them are tough enough for the real stuff: good genes.

How might Zahavi's principle apply to songs? A mockingbird that sings a hundred and fifty different imitations of all its territory's other birds, combined and contrasted in an intricate musical mix, or a brown thrasher that lets loose two thousand distinct strophes in its inexhaustible repertoire is spending minutes, even hours expending a huge amount of energy. Wow, thinks the female, this guy can really keep going, he's got all the energy I need, even if he wastes all that time singing. Her individual choice serves to further this species on into the future.

Is extensive singing actually costly to birds? It is well-known that in breeding season nightingales do not eat at night, but only sing. Robert Thomas of the University of Bristol in England measured how much eight individually banded male nightingales weighed before dusk, at the time of their final feeding before the night's performances, then again at dawn, after the singing calmed down. Like most other songbirds, these nightingales stopped foraging at least an hour before dusk, and did not start again until nearly an hour and a half after dawn. During their long bouts of song they tended not to move around much, suggesting to Thomas that the nightingale is basically a daytime bird, except for its special talent for nocturnal singing during these few weeks in early spring.

The birds that sang more lost more weight during the night. The birds who bulked up with more food the previous day gave more extensive nighttime performances. The informal conclusion? Singing takes a lot out of nightingales, so if you're preparing them for a heavy night of treetop music—more than forty minutes of song per hour until dawn—then you had best encourage them to stock up on worms to get ready. And in the morning, they sure will be hungry. The scientific conclusion? Singing has important metabolic costs. Music for the nightingale is not cheap.

Previous studies on other birds have not been as conclusive—the familiar *cockadoodledo,* for example, isn't hard for a rooster to muster each dawn. But for a nightingale's song to be considered a risk, a handicap, or a genuine biological excess, science needs data to show that it ain't easy. Thomas believes his bird mass measurements do exactly this, and he also suggests other factors that make song risky that are important but probably harder to measure, such as predation costs. When nightingales hear the calls of tawny owls, their major predator, they tend to quiet down, but they do keep singing. Any sound they make might alert an owl to swoop down and tear the nightingale from the branch. That would be one expensive song.

It took weeks of analysis of video recordings for Thomas to accurately collect data. Did he examine enough birds for long enough? Only eight. Critics of his work would demand a larger sample. Do the same experiment with eighty birds for several seasons. That would take a lot more time. Would the result be more conclusive or more ambiguous? It is not likely anyone will duplicate the experiment, as the work would be difficult and redundant. How many nightingale songs ought to be weighed?

These questions veer far from the reasons why birds sing. I am amazed and impressed by the diligence of science, but for all the hours Thomas spent poring over his videotapes, his report offers more pages of calculation than observation. I care less about how much the nightingales were singing than *what* they were singing. Was it interesting? How did it feel to sit in the dark and listen to it? I don't want to know how much it costs. The Zahavi hypothesis seems an interesting one to test, but it makes song seem only important in an incidental way; I don't want to consider music, or anything beautiful, to be a handicap. Science can still listen more closely before it retreats into the safety of measurement. This pushes me into what Richard Dawkins calls "the argument from personal disbelief." Why should my hunches have anything to do with what's true? In the end, we'll realize we are much closer to birds than we at first want to admit. Our intuitions can help us make sense of the way other species live, but we should not be afraid of what careful experiments conclude.

DARWIN'S HEIRS have sought explanations for what he was content to admire as beautiful. They have gone quite far through generations of theorizing. But there is another line of descendants to consider: the great naturalists who specialized in observation and description. Armed with knowledge of natural selection, they were able to see farther than ever before.

John Burroughs, great chronicler of the natural world in my own Hudson Valley and Catskill Mountains, wrote in *Ways of Nature* a hundred years ago of a regular weekend visitor from the city who asked a country bird man if he could take her to hear the song of the bluebird. "What, you have never heard the bluebird, in all your years here?" said the astonished ornithologist. "No," said the impatient woman, time being of the essence. "Then," he shook his head, "you will never hear it."

Bird song is something you must reach out to love before it will reveal itself to you. I don't think Burroughs or any other of our pioneer naturalists would have been satisfied with the handicap principle. It almost mocks the beauty of nature that they praised in celebration. Bird song is not immediately something to analyze or count but a pull on the senses—not only for us, but also for the birds. If we do not reach out to it and imprint it on our memories, we will never hear it, and if birds do not embrace and feel it, it will mean nothing to them. "The song is not all in the singing, any more than the wit is all in the saying. It is in the occasion, the surroundings, the spirit of which it is the expression."

The response of nineteenth-century naturalists to bird song falls somewhere in the valley between science and poetry. A naturalist spends long hours paying complete attention to nature, not to conduct experiments or derive a satisfaction in certainty, but for the sheer pleasure of awareness. To look, listen, and learn, and to come home with that kind of knowledge that is closer to acquaintance, to gain a familiarity with the world of birds, to share in their experience, not necessarily to explain it. Here is how Wilson Flagg, author of *A Year with the Birds* (1881), elaborates on Thomas's question: Why do some birds sing at night when it clearly exposes them to the dangers of the night?

> Why do they take pleasure in singing when no one will come in answer to their call? Have they their worship like religious beings; and are their midnight lays but the fervent outpourings of their devotions? Do they rejoice like the clouds in the presence of the moon, hailing her beams as a pleasant relief from the darkness that has surrounded them? Or, in the silence of the night, are their songs but responses to the sounds of the trees, when they bow their heads and shake their rustling leaves to the wind? When they listen to the streamlet that makes audible melody in the hush of the night, do they not answer to it from their leafy perch?

The rhapsody tumbles on through layers of further queries. I suppose each of these factors could be tested with a scrupulous scientific experiment. Turn off the rushing cascade and see if the bird still sings.

Take him away from the moon. What kind of song comes out then? As we shall see, far stranger experiments than these have been tried and learned from. But the yearning reach of dreamy woodland naturalists has not really been exceeded. Through speculation, they dared to make these songs matter. Since no useful purpose can easily explain why night birds risk their lives when they put their songs out on a limb, Flagg will hazard a wild ungainly guess: the song is there "to soften the sadness of the dark." For its own sake. The darkness must be complete, containing birds as well as people, the whole of the world.

When nature writing was enjoying its greatest popularity, scientists of the nineteenth century were in many ways at the same point we are today. "The majority of ornithologists agree in ascribing an erotic character to the songs of birds; not only the melting melodies, but also those of their tones that are discordant to the human ear," writes Dr. B. Placzeck in *Popular Science* in the 1880s. "All are regarded as love-notes." It sounds a bit more alluring than the language scientists permit themselves today. But Placzeck is not satisfied with the functional explanation. He continues, "The bird sings, to a large extent, for his own pleasure; for he frequently lets himself out lustily when he knows he is all alone."

In *Birds in the Bush* (1893) naturalist Bradford Torrey confronts the first robin of spring engaged in song, perched in a leafless tree for all to see. Rather than asking and musing why the bird sings, he imagines how the robin might respond: Hey, you human down there, you think I should have to *explain* why I sing? You seem to have forgotten that everybody sings. If bees, crickets, and mosquitoes all sing without anyone bothering them about it, why should I, a robin, get the brunt of such wondering? And how about you? Why indeed do you sing yourself? Torrey knew that some birds learned their song while others were born with the ability to sing. He also knew that knowing this doesn't shed much light on the fundamental question. Although suspicious of the idea that there is "any real connection between moral character and the possession of wings," he continued, as most of us do, to somehow admire birds' ability to provide the world with joyous song throughout the day, even though there is so much else they must do to survive. And not least, "possibly their habit of saluting the rising and setting sun might be the first glimmerings of original religion."

The urge to describe if not to decode beauty always stays with us. Poring through old and new accounts of birds at work and play, I seek desperately for some attempt to link the rational and the rhythmic, the indifferent and the essential, the magic and the meticulous, in the investigation of bird song. The more we advance in deciphering the natural world, the less irony seems to appear in the literature. Even the possibility of endless wonder is hushed up. Meanwhile, skeptics huddle, muttering that the most important things in life can never be completely explained. The two camps rarely speak to each other.

There have been rare hybrids, unusual individuals who don't fall easily into either camp. Darwin was one, with his applause for birds' natural and native aesthetic sense. There are others who dared to think outside the boxes of poet, scientist, or musician, finding interesting ways to blend these distinct perspectives. Walter Garstang, professor of zoology at the University of Leeds in England in the early twentieth century, is most known to science for a paper that demolished the recapitulation theory in evolution. But he is best known to the general public for writing one of the most unusual of all bird song books, *Songs of the Birds* (1922), a work that walks the line between science and poetry.

The first half of Garstang's book is a call for a new science of bird sound. Sound is difficult to analyze because time cannot be stopped. "Our object is to study the expression of emotion in sound. Our difficulty is that we are invading a completely immaterial world of things that vanish as soon as they come into being." Not even considering that sound recording technology might help him, Garstang was an unusually fastidious listener to the abstractions of bird music, and he was able to link what he heard with Darwin's ideas of the differentiation of bird species and the way song might illustrate evolutionary connections among them. Here is his insightful account of hearing the nightingale's song as a mark of the bird's place in the scheme of natural selection:

> The long-drawn crescendo wail, which cuts through his sequences at intervals with a plaintive falling cadence, seems to be the exact counterpart of the alarm-rattle intermezzo of his larger thrush cousins, though all signs of the chuckle have gone, and the chime has been smoothed into a nearly continuous slur of arresting beauty. Notice the art in this achievement. The essence of the nightingale's song is insistent repetition of highly resonant monosyllabic notes at contrasted levels of pitch. The most perfect foil to this is unbroken continuity of purely musical tones right through the scale. He attains it by transfiguring an heirloom which was originally the most essentially iterative and resonant of all the cries at his disposal! What a stir in the coverts the bold upstart must have created who gave the final touch to this transformation of the family rattle!

This is a most poetic kind of science. Here's a listener who is actually trying to tackle the aesthetics of evolution with head-on analysis. Garstang proceeds to demand some kind of accurate notation of bird songs. He did not like the musical notations his predecessors had used, and instead pioneered a kind of abstract syllabic notation that seems to descend from the *jug jug tereo* of the Elizabethan age and the *cheer-up cheer-ups* of John Clare. The pursuit of this grail pushes scientist Garstang in a very unusual direction.

Garstang was lured in by two goals: first, accurate representation of the structure and timbre of bird songs, for which he decided strange and recognizable words were the way to go. Then, he wished to enter the heart of the song itself. Science, he writes, may try to trace the origin of song or aim to discover its real purpose, but the song itself is still an expression of the bird's emotion, a "prolonged elevation of the spirit," a transformation of his everyday squawks and peeps into something higher, "an expression of the whole joy of life at its climax of well-being." It is this insight that leads Garstang into poetry.

The second half of his book is composed of his own rather optimistic poems based on these nuanced transcriptions of bird songs, mixed in with a light verse that celebrates all the wonderful ways birds can make us feel. These were published in various English newspapers, and although never celebrated as great poetry, they seem to have been quite popular during their time. Here is Garstang's account of a skylark heralding the dawn:

Swee! Swee! Swee! Swee!
Zwée-o! Zwée-o! Zwée-o! Zwée-o!
Sís-is-is-Swée! Sís-is-is-Swée!
Joo! Joo! Joo! Joo!
Jée-o! Jée-o! Síssy-sejóo!
Jít!, jit!, Jít!, jit!, Jít!, jit! Dzóo!
Zée! Wee, wée, wee! Sís-is-is-Swée!
Swée-o, Swée! Swée-o, Swée!
Swée, swee, swée, swee, swée, swee! Swée!

Garstang didn't think he was playing around. He considered these new words to be a tool of science: a rigorous, exact transcription. When a newspaper critic asked if "this Czechoslovakian poetry" was really necessary, Garstang answered that the deciphering of new languages requires new syllables. That's why he put them in italics. "Old Sam Peabody" is not enough.

Here is a scientist who recognizes that the subject of his inquiry tests the limit of reason. He became a poet in order to find something genuine to say about joy. His poems may be noticed today only as a footnote to the inscrutability of the world of birds, but it cannot be denied that they express passion and commitment to the sounds pouring forth from the trees. Here's what Garstang heard in the flight song of the tree pipit:

Wait now, watch, and keep quite still!
Listen to the voiceless trill
Coming from that slender bill
And charged with joy unspeakable of his devotion!
The thrill of rapture waxes, wanes, and dies
As he alights again upon his tree;
And then, if all goes well, there will arise:

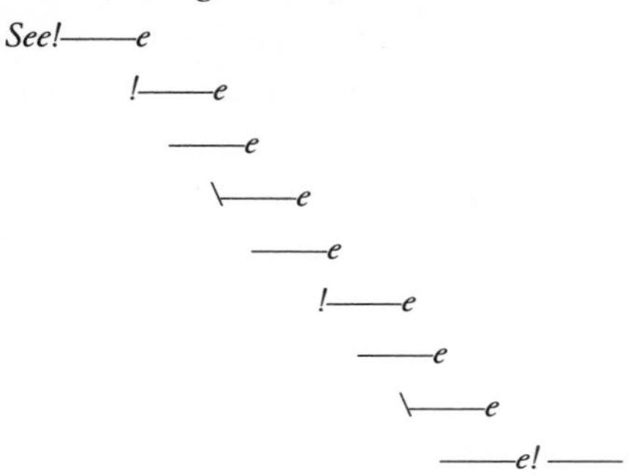

A sign of esctasy
So sweet and infinite. . . .
A whisper to the sunset sky,
Which, softly sweeping down the scale of music's ocean
Brings answering echoes from the wells of deep emotion.

Such a combination of attentiveness and passion is indeed rare. If anything, it shows how hard it is to make science and poetry one and the same. But maybe "Why do birds sing?" is not a good scientific question. I admire Garstang for trying, and I wish others would take such risks. In our own era, with all the tools of computer transcription, statistics, and scientific advance, is science still willing to admit it needs poetry to grasp the full meaning of bird song? Garstang wanted to share in the exaltation of larks. As he transcribed their music into wild and crazy alien syllables, he couldn't help but be carried away.

Garstang's wild poetry seems less like science and more like dadaism. It is reminiscent of the famous *Ursonate,* the pioneering sound art piece composed by Kurt Schwitters, otherwise most known for his collages and paintings. This thirty-minute sound poem, begun in 1922 and refined over the whole next decade, is composed of what some would call nonsense syllables, but bird listeners would quickly hear how Schwitters was influenced by the mnemonic notation used to humanize bird song into something we might remember. Here is an excerpt:

Ooobee tatta tuu
Ooobee tatta tuu
Ooobee tatta tuii Ee
Ooobee tatta tuii Ee
Ooobee tatta tuiiEe tuiiEe
Ooobee tatta tuiiEe tuiiEe

Tatta tatta tuiiEe tuiiEe
Tatta tatta tuiiEe tuiiEe

Tilla lalla tilla lalla
Tilla lalla tilla lalla

Tuii tuii tuii tuii
Tuii tuii tuii tuii

Tee tee tee tee
Tee tee tee tee

Schwitters is playing with sound purely for its own sake, but I am struck by the resemblance to the supposedly scientific poetry of Garstang. Both pushed the bounds of their respective disciplines to stretch language toward the limits of syntax and on to the noise of the world. The babel of birds awakens us all.

GARSTANG THOUGHT THAT A POETRY of new words was needed to articulate the raw feeling of a bird's song. Another pioneer thought that musical notation, however human and imperfect, was the way to prove the inherent musicality of avian sounds, because notation is the most accurate way we humans have to write down rhythm, pitch, and form, although it has a harder time with timbre or tone quality. The work of American naturalist and educator F. Schuyler Mathews straddles a similar fence between science and music. In 1907, then with a second edition in 1921—reprinted often because people love the audacity of it—Mathews offered his *Field Book of Wild Birds and Their Music,* which begins with an intuition that seems obvious to the listener not yet tainted with the mechanistic view of extreme Darwinism—"The bird sings first for love of music, and second for love of the lady . . . ." Mathews advises us to look for the cause of bird song in the internal life of the bird rather than in the comfort of purpose. The way inside the bird, Mathews believed, was to grasp for the song as if it were music.

Why aren't charming syllables enough? Look, he says. The white-throated sparrow is supposed to sing "Old Sam Peabody, Peabody, Peabody," but if you listen closely, he speaks his coda in a specifically uneven way: *peeee bu-di peeee bu-di peeee bu-di* in a swinging, jazzy uneven rhythm, which Mathews says language cannot render clearly. Musical notes are needed. Mathews considered bird song to be music in miniature, each song a tiny vignette of rhythm, form, and phrase that begins to make sense but does not quite succeed.

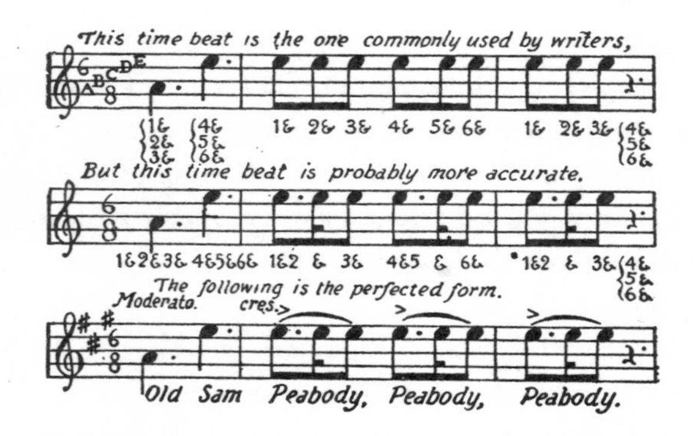

F. S. MATHEWS' WHITE-THROATED SPARROW MUSIC

He transcribes and retranscribes the twitters of a more complex avian tune—that of the song sparrow, with the hope that the bird will utter a phrase complete enough to his liking:

F. S. MATHEWS' SONG SPARROW MUSIC

"Why not go on?" Mathews implores the bird. As much as he loves what he hears, he does not quite hear what he wants. Throughout the book he chastises our feathered friends for refusing to end their haphazard phrases with the proper cadences. "The fact is, the bird has not *arrived;* there is still no point to his song. He makes a fine start, but he nearly always fails to finish on the tonic, or, for that matter, anywhere at all."

Rather a strong condemnation for a book that proceeds to devote nearly three hundred pages to these supposedly pointless songs. Mathews clearly has a love–hate relationship with his subjects. When he enters the strangeness of their music, he too comes to realize the limits of notation's ability to make sense of these sounds. See what he has to say about the alien music of the screech owl, one of the most fearsome sounds of the forest night:

> Their notes were simply weird, a sort of cross between a sneeze and the wheeze of a pair of leathern bellows with the wail of a half-frozen

puppy thrown in to make matters more mysterious. I shortly came to the conclusion that these were young birds which had not yet learned to sing properly, so I gave them a lesson or two, at the same time profiting by the experience, and getting a few lessons for myself.

F. S. MATHEWS' SCREECH OWL MUSIC

The song can be easily represented within our familiar musical notation, but it contains a free, searching quality that most set melodies seem to eliminate the moment they reach closure or cadence. Perhaps nature isn't full of stories or structures waiting to be told. It sings of an endlessness and recurrence that flies in the face of our imagined need to say that music is a statement in sound with clear beginnings, middles, and ends. Start listening to bird song as music and the borders of music itself break open. Mathews saw this when he realized that some of his favorite bird songs lay at the limits of what notation could express. The rhythmic herald of the bobolink stumped him as he tried to trace where the warbles were going:

F. S. MATHEWS' BOBOLINK MUSIC

Just as Garstang's weird words anticipated Schwitters' dada with pages of birdlike syllables and rhythms in language that speaks but does not say, Mathews stretches musical notation to the brink of the avant-garde. This transcription of a meadow bird looks like a score from Stockhausen! The more Mathews listened and wrote down, the closer he came to the music of the future.

Mathews preferred clear and simple songs with beginnings, middles, and ends—the predictable sort of bird music. But he may have discovered why the harder songs always sound sudden and fresh—we cannot quite grasp or decode them. Maybe Mathews' complaint that the bobolinks' melody "never arrives anywhere" could be translated to "they have nothing to prove, they never know ennui."

Mathews is frustrated by so many of the familiar and simpler songs he is able to turn into musical notation, even with piano accompaniment ("solely to demonstrate the musical content, not for identification purposes"), but the one song he admires most is tough to turn into graceful notation. It belongs to the elusive gray-brown hermit thrush, whose rising double-fluted cadence is often judged to be the finest North American bird timbre. It does not look like much when he writes it out, because the notation cannot contain all the song says to him:

F. S. MATHEWS' HERMIT THRUSH MUSIC

This is what sets Mathews' spirit to soar: a single motif, a warbling rise. Standard notation cannot express the depth of tone or endlessly varying quality of the phrases. Mathews' written response to the hermit thrush's song is one of Garstang's legacies, the first of several we shall find. It is a moment when an observer who is trying his best to be objective is suddenly driven into poetry at the sheer magnificence of the sound of a bird:

> Bird songs are most ethereal things, a great deal like the wonderful tinting and delicate spiral weaving in Venetian glass. . . . How can the meagre outlines of music notation convey such truths! Who can justly report the Hermit's song! There is a silvery sustained tone like that of a flute, then a burst of brilliant scintillating music . . .
>
> and the song's complete
> With such a wealth of melody sweet
> As never the organ pipe could blow
> And never musician think or know!

No accident. Bird song *makes* its most attentive human listeners surge into poetry. The quality of the poem matters less than the need to go there: we want to stretch language beyond its ability to explain, into its chance to evoke.

Think of the famous Navajo blessing, first as some wild syllables that look alien to us only because we do not know the language. Once we translate it, they offer an exaltation of the world that can hardly be improved:

Hózhóogo naasháa doo
Shitsijí' hózhóogo naasháa doo
Shikéédéé hózhóogo naasháa doo
Shideigi hózhóogo naasháa doo
T'áá altso shinaagóó hózhóogo naasháa doo
Hózhó náhásdlíí'
Hózhó náhásdlíí'
Hózhó náhásdlíí'
Hózhó náhásdlíí'

In beauty I walk
With beauty before me I walk

With beauty behind me I walk
With beauty above me I walk
With beauty around me I walk
It has become beauty again
It has become beauty again
It has become beauty again
It has become beauty again

The world is blessed and birds are in it. We witness the *hózhó* of the song whether we try to write it down, learn from it, or just use it to enter the way of the bird from within. Musicians, amazed by the resilience of bird song, transcribe the oldest and newest music they know. When poets try for a glimmer of that great meaning, they hear either unbridled passion or feel a deep longing to fit into a nature that has made us too self-conscious to bask in its perfection. When science tries to figure the purpose of all this natural music, it gives us the weight in grams of one bird's song.

Birds may hear beauty in our alien sounds as well. In southeastern Australia they tell a story about the superb lyrebird, widespread in the hills of the states of Victoria and New South Wales. This species possesses a shared and learned sense of song that is passed down from generation to generation. In the 1930s in Dorrigo, New South Wales, a flute-playing farmer kept a young lyrebird as a pet for several years. In all that time, the bird learned to imitate just one small fragment of the farmer's flute playing. He had no use for any other human sound. After a time, the farmer released the bird into the forest.

Thirty years later, lyrebirds in the adjacent New England National Park were found to have flutelike elements in their song, a sound not heard in other populations of superb lyrebirds. Further analysis of the song showed that the phrase contained elements of two popular tunes of the 1930s, "Mosquito Dance" and "The Keel Row." As lyrebirds can sing two melodies simultaneously, through several generations this population had created its own distinctive territorial song blending the two melodies into a single compressed phrase, refining it from generation to generation.

It is now seventy years since a lyrebird learned these fragments, and today the flute song has been heard a hundred kilometers from the original source. A human tune is spreading through the lyrebird world, as they've decided through generations to prefer just two shards of our particular music. Beauty goes on through habit, not accident. In beauty we walk and in beauty we listen. The birds too fly with it and preserve it in sound. Does nature have a good reason for holding on to some melodies few humans remember? We may revel in this discovery or find new tools to further dissect the source, freezing on paper the fleeting wonders sung out in moments of time.

CHAPTER 4

The Song Machine

THE REMARKABLE ALBERT's and superb lyrebirds combine virtuosic powers of imitation with outlandish display tactics. They are extreme examples of sexual selection and the peculiar adaptations of the bird world. Yet lyrebirds were largely unknown to the world until the Australian Broadcasting Commission beamed their songs live on the airwaves in 1932. Thousands tuned in to hear the drama of this unique bird's song, and the world was alerted to its beauty. People realized the importance of preserving lyrebird habitat so that this ultimate bird song might endure, and the species soon gained official protection.

Before the advent of recording technology, listeners to bird song had to trust their own ears. Sound recording was invented by Edison in the last decades of the nineteenth century, and over the years it was gradually improved and then combined with radio to make rare sounds available all over the world. Before television nature programs there were of course radio nature programs, and bird songs were often featured. They were one of the first aspects of the natural world to touch millions of listeners through the new wireless technology.

With time, sound recording moved through the wax cylinder and the pressed record to the magnetic tape recorder, which offered new flexibility to field recordists. During World War II the sonograph was invented, a device that could visually record the details of sound—pitch, duration, and rhythm—without relying on fallible human ears. This technology enabled scientists to see the truly alien organizational principles at work in bird sounds: a huge amount of information produced per second, nonharmonic sounds, irregular rhythms moving by at a speed far too fast for us to grasp. The sonograph made the analysis of bird song far more objective, closer to geometry than to poetry.

Without recordings, investigators of bird song had to be expert listeners and observers, training their senses to follow what they heard. There was no way to be objective about sounds that seemed so different to each listener. One man's *zwee* could be another's *sree*. One country's "Peabody, Peabody, Peabody" could be "Canada, Canada, Canada" just over the border. And of course one man's music is another's noise. Could machines bring us beyond personal opinion?

Before "recording" meant preserving a performance, either of human or bird musicians, it meant learning. And learning is what ornithologists began to use bird sound tapes for. Ludwig Koch was the first celebrity among natural sound recordists. Like many pioneers of new technologies, he was prone to grandiose statements about himself, some of which are probably true. He claims to have been the first person *ever* to record the sound of a bird—as an eight-year-old boy in 1889, when he recorded onto an Edison wax cylinder the sound of a white-rumped shama, that great Asian singer I heard in the aviary. By the 1920s he was working for the German company EMI, and he came up with the idea for the first "sound book" on bird songs accompanied by gramophone records.

Simply describing bird songs on paper was not enough to teach us. "Those musical notations and curves," he wrote in his *Memoirs of a Birdman,* "mean nothing either to a scientist or to a bird-lover." They convey none of the great sweetness of sound real birds in the wilds let loose. Koch was also one of the first to notice that caged birds deprived of their natural upbringing often fail to produce their species' characteristic sounds. We need, he announced, to go out into the wild and bring the real sounds back inside.

In the early days of recording this was no mean feat. Equipment was heavy and cumbersome to lug around. Once the recordings were brought back into the studio and made available for all to hear, the interest in bird songs was tremendous in his native Germany, though the rising leader Adolf Hitler preferred recordings of the roars of German stags. Later in the thirties Koch quietly made his way across the border to Switzerland and then to London, where he was welcomed into the close-knit world of British bird lovers and given great opportunities by the BBC to develop bird song programs that became famous as far as the voice of the empire was heard.

Koch writes of his encounters with royalty, statesmen, and the common man, all of whom profess a great love for the songs of birds and especially Koch's collection thereof. There are harrowing journeys to remote places with ungainly equipment to catch the elusive sounds of rare and shy species. He described the hunt for the hard-to-hear burble of the female cuckoo, a sound only made when the bird is in fast flight:

> I was standing close to my recording equipment, with headphones on, when I saw the two cuckoos, about half a mile away, racing in the direction of the microphone, and I saw the engineer begin cutting as I gave him the sign, "Cut!" And in this very fraction of a second the female cuckoo uttered her bubbling note so close to the edge of the disc—which was not wax but aluminum acetate—that we had a great deal of trouble to get a good dubbing in the BBC studios. This shows that in spite of my observing the cuckoos at a great distance their speed was such that the cutting remained the greatest gamble, and I could understand Mr. Lees remarking to me, "you missed it again."

Koch, the David Attenborough of the radio era, seems to have always got his song in the end. "Call me a fanatic," he ends his account, "I am a fanatic. . . . But I like to remember that I share that fanaticism with one infinitely greater than myself who, had recording gear and the radio been available to him, might have done the same thing while transmuting his close relations with birds for a divine and humanitarian purpose: St. Francis of Assisi." You have to admire a man who compares himself to a saint. Koch recognized the power of the new technology to make him and the birds celebrities together.

And he was instrumental in setting the science of bird song in motion. Sitting in his book-lined office at the University of California at Davis, Peter Marler, probably the greatest bird song scientist alive today, describes how he got started:

> I really began as an amateur bird watcher, back in high school. One day I went to the London Zoological Society to hear Ludwig Koch present his work in a strong Austro-German accent, very endearing. I was absolutely thrilled. He was especially interested in chaffinch song, and somehow that was an *a ha* experience for me. I'm not sure if I had already begun transcribing chaffinch song, but by then or soon after I developed this primitive script, long before I knew anything of sonograms or tape recorders.

Marler has lived and worked through all those decades when technology radically changed our ability to make sense of sound, and his recollections will illuminate many of the stories that follow.

With the ability to listen to the same recorded sound over and over, it became possible to create more precise ways of getting bird songs onto the page. The most detailed of these efforts in the mid-twentieth century was Aretas Saunders's *Guide to Bird Songs*, first published in 1935. Saunders believed that a precise method of writing the sounds was needed that did not presuppose the bird watchers' ability to read musical notation, since that is a specialized skill and everyone ought to be able to read bird song notation. He did not want to promote what he saw as an erroneous tendency to link bird song too closely with human music. His notation is presented as an exact, scientific record, but to me it looks more like secret diagrams or conceptual art.

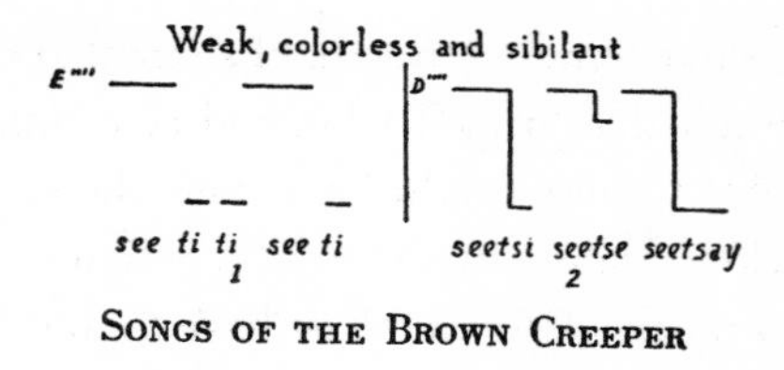

SONGS OF THE BROWN CREEPER

Like Mathews a generation before him, Saunders sometimes indicates a bit of disdain for his subjects. The brown creeper is a little inconspicuous brown bird who circles up tree trunks in a rising helix. "Sibilant"? Not much of a song perhaps, but does the notation capture it? It is the sort of graph that Koch said "means nothing." The lines together with the words do seem suggestive to me.

Here is Saunders's account of my favorite, the indefatigable mockingbird, whose song we will decipher more deeply in Chapter 8. These lines begin to reflect the rhythm and the flow of this shimmering song. Just a piece of it, but perhaps enough for you to identify the song should you hear it and have the book in hand. That is Saunders's primary aim here, to enable bird watchers to become bird listeners, learning to distinguish a species by its sounds. Today we have tapes, CDs, DVDs, and digital mp3 players with thousands of sounds to check in the field. Fifty years ago we had only diagrams and transcriptions. Recordings had to be heard at home.

Wavy curves and diagrams, musical notations, churring languages—which is the best way to represent the song of a bird? They are all creative, subjective, personal. Good for you if you want to note down *something* in order to remember what a bird sounds like. But idiosyncratic notation is not so useful if you want to tell others precisely what sounds a bird is making.

Enter the machine. In the 1940s researchers at Bell Telephone Laboratories in New Jersey invented the sound spectrograph, otherwise known as the sonograph, at first for the purpose of identifying possible criminals by their voiceprints, a printout as distinctive as a fingerprint. The device, a result of secret technology developed during World War II, prints out onto paper a sound spectrogram (or *sono-*

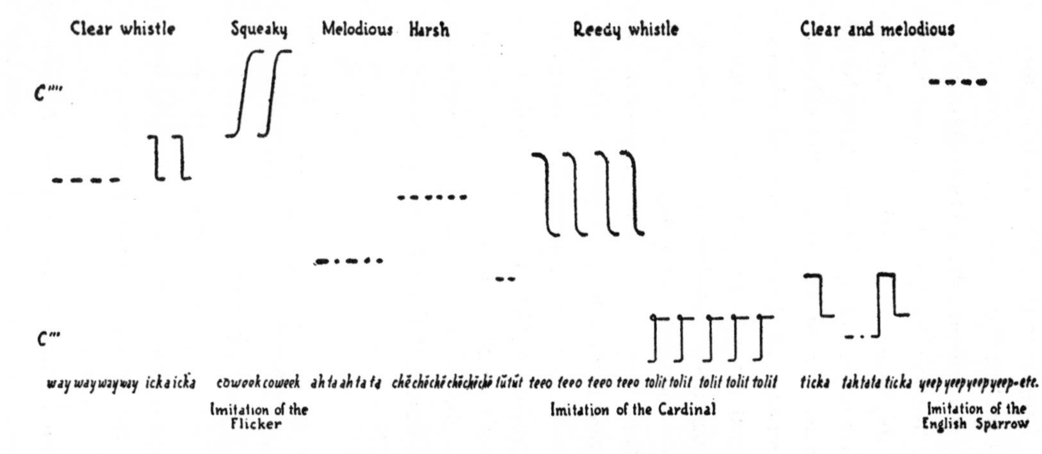

PORTION OF A SONG OF THE MOCKINGBIRD

The time between groups of notes has been shortened in order to get more of the record on a page. The singing of these phrases occupied about twenty seconds.

gram), which is a graph of the two most important variables of sound: frequency versus time.

W. H. Thorpe, founder of the zoological laboratory at Cambridge University, immediately saw the sonograph's potential for the analysis of bird songs. His star student, Peter Marler, remembers the day the machine arrived. "It was clear from the moment that we opened up the package that this would be something important. In 1949 there was only one other sonograph in Britain, at the Naval Laboratory, where they were using it for underwater surveillance. Thorpe was lucky enough to get one—we knew at once that it had almost magical potential." At last, the strict details of the song could be objectively rendered on the page, and we could approach an exactness far beyond mnemonic musings and quirky squiggles.

First Marler and Thorpe had to learn to read the odd marks the machine printed out. If the sound is a simple scale of clear, whistled notes, like the innate call of the pileated (or little) tinamou of Costa Rica, you see a series of simple rising lines. The flat lines mean simple pitches, like that of a single flute tone. It's a simple, very slow scale moving upward, quite rare in the avian world. When we hear such a song, it is easily as beautiful as music.

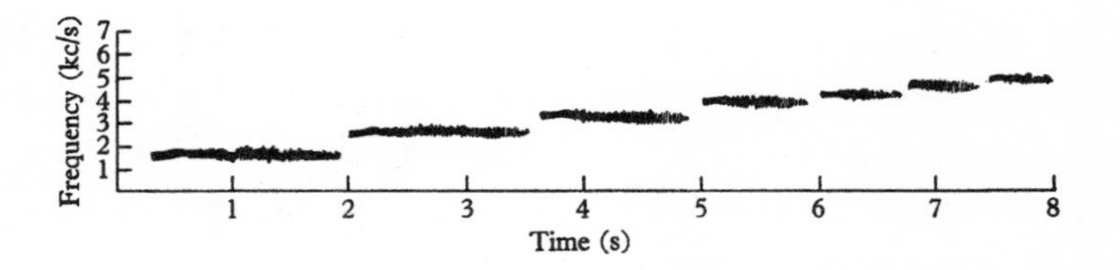

W. H. THORPE'S SONOGRAM OF THE RISING SCALE OF THE PILEATED TINAMOU

Whistle-type bird songs produce sonograms that are easiest to compare with musical representation. Here is the eastern wood pewee. Notice that the sonogram reveals more subtle nuance in the shape of the phrase. And with the layering of one frequency upon another, they are able to depict the timbre of a bird song in a way musical notation cannot. Thorpe admires this music in miniature, praising its "well-balanced line." In Chapter 6 we will learn why ethologist Wallace Craig wrote a two hundred-page book called *The Song of the Wood Pewee,* the most detailed study ever attempted on a single bird song.

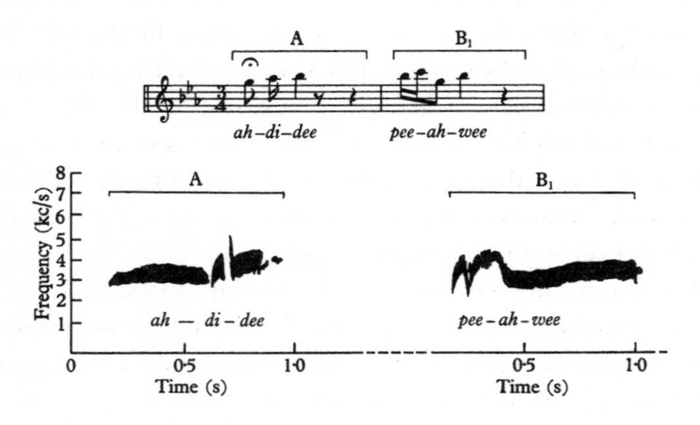

W. H. THORPE'S COMPARISON OF MUSICAL AND SONOGRAPHIC NOTATION OF THE SONG OF THE EASTERN WOOD PEWEE

There are many sonograms in the pages that follow, not because I expect you to decipher them, but mostly because I find them to be quite beautiful. Especially the vintage ones from the fifties and sixties, when scientists had to ink them by hand on tracing paper over the original messy machine printouts. The most intricate I've seen from this era is the Gouldian finch:

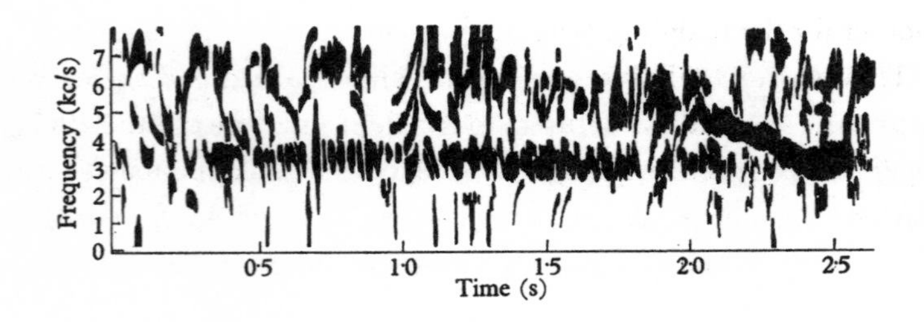

A sonogram of such complexity indicates that three different sounds are being produced at the same time, two overlapping chirrups above the trace of a drone, revealing the awesome complexity of the syrinx, the birds' vocal organ, which is far more flexible than the human larynx. Sonograms today are instantly spit out by a computer, using numerous easy-to-use software programs, some of which are available for free. There is tremendous flexibility in how they can be made to look, but they lack the emphatic brushstroke appearance of the earlier sonograms.

You might think I am bringing unnecessary aestheticism to the evaluation of something that is meant to be scientific and objective. But consider what these diagrams are meant to show: the horizontal (time) and vertical (pitch, timbre) qualities of sound. Interpreting them takes practice—in listening and looking. Sonograms may be used to show what our hearing is not fine enough to discriminate: how a bird learns its song, shares it with other members of its species, and sings slightly different songs from its neighbors.

Thorpe's most influential work was done on the chaffinch, a common European bird. At first he thought the new technology should be tried out on the bird's simpler sounds, not the songs but the calls. He set his student to work on this project, and Marler wrote one of his first scientific papers to demonstrate that the calls of the chaffinch had very specific meanings. The paper was published under the provocative title "The voice of the chaffinch and its function as a language." Marler found twelve specific calls, all innate. The flight call, the social call, the injury call, the aggressive call, the *tew* alarm call, the *seee* alarm call, and the *huit* alarm call. Three courtship calls, *kseep, tchirp,* and *seep*. One call for the nestling begging for food, and another for the fledgling begging for food.

This was in 1955. No one had previously imagined the calls of birds to be so tightly structured. How did people react to this knowledge? "I don't think anyone paid all that much attention to it," shrugs Marler:

> It's not like Watson and Crick figuring out the genetic code, with blazing headlines everywhere. It was a rather lonely enterprise. It was a new thing to do; it had not been possible until that time to begin thinking about such issues in anything other than intuitive terms which had already become the targets of skepticism and distrust. I was so enthusiastic it didn't bother me in the slightest, though my enthusiasm was not immediately contagious.

If you are seeking specific meanings in bird sounds, there is much more to be said about calls than about songs. Marler later discovered that the *seee* alarm call is constant over several different species of birds. On page 67 are sonograms of calls given by four different British birds when a hawk flies overhead, all sounding something like a descending *seeeee . . . !* Looking at this group of sonograms, I feel a chill in my heart. They even resemble the swoop of a big, dangerous bird shadowing across the sky. Such remarkable trans-species similarity has led later researchers to wonder if there might not be a universal language of specific distress that animals all know. Astonishing perhaps, but also practical. Any bird in the vicinity ought to know if a hawk is flying over. Why not make it easy for them to share this information?

Marler laughs now that he thought to call bird calls a "language" in his youthful days. But around the same time, Thorpe noted that "however great the gulf which divides animal from human language, there is no single characteristic which can be used as an infallible criterion for distinguishing bird from human language." Nearly all the qualities of human speech are found somewhere in the animal world; what is unique is the way we combine them to achieve the greatest flexibility of expression.

Bird calls, with their decipherable and specific meaning, do seem like a simple language for conveying precise information. So the Netsilik in Greenland say that in the Beginning, "people and animals spoke the same language. That was the time when words were like magic." Not anymore, but in ancient times, we could understand

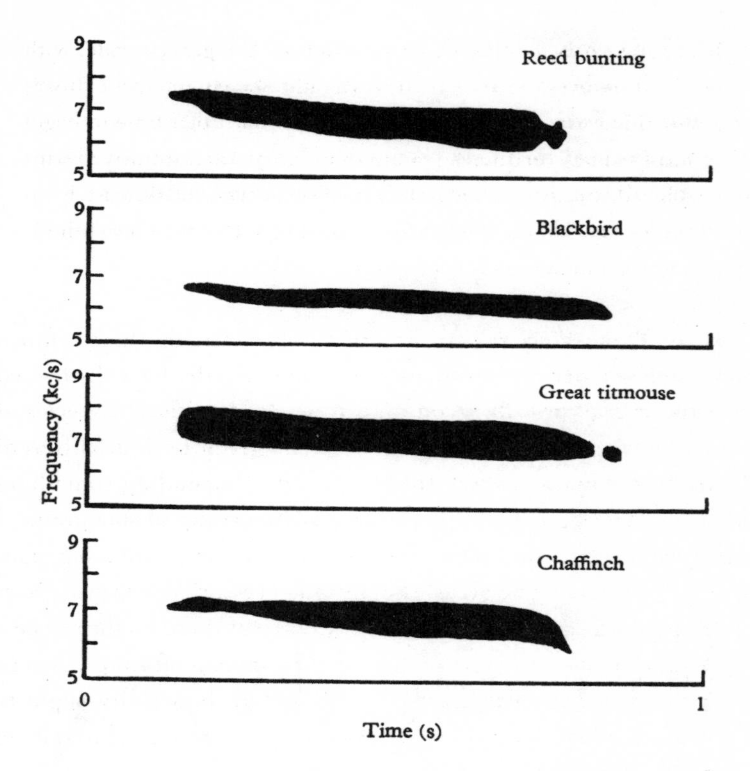

FOUR BRITISH BIRDS MAKE THIS SIMILAR CALL ONLY WHEN A HAWK FLIES OVERHEAD

what the birds were saying. Thorpe and Marler used the magic of the sonograph to decode the call sounds of the chaffinch, and now we once again know what they're saying.

But the Netsilik and the scientists both know there is more to the story: the most lyrical sounds produced by both birds and humans are interesting in themselves alone. Their specific meaning is less clear, but we love them all the more for that. We call our sounds music, we call the bird's noises songs. The information they convey is not easily decoded. They are usually more elaborate and self-contained as expressions; more composed, more extensive, more beautiful.

The song of the chaffinch was perfect for Thorpe to feed into the new machine because it is both musical and concise, with an exact recognizable form. There is enough complexity in it to have kept scientists transfixed for many years since. But even with the gleaming new machine in his lab, when Thorpe wanted to express what it sounds like, he calls up Walter Garstang's mnemonic words:

phrase 1a: *chip-chip-chip-chip*
phrase 1b: *tell-tell-tell-tell*
phrase 2 *cherry-erry-erry-erry*
phrase 3a-b *tissy-che-wee-ooo*

Here's how it looks in a sonogram:

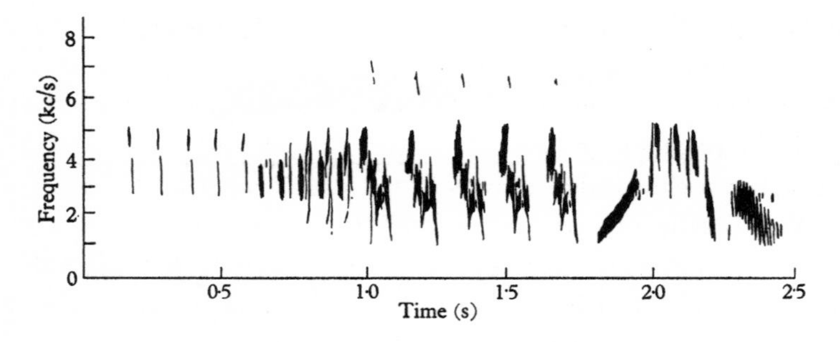

THORPE'S SONOGRAM OF A CHAFFINCH SONG

There it is, Garstang's verbal flourish and Saunders's squiggly lines combined into something more scientific and more lovely, an elegant drawing. This is the inked-over printout that so captivated Marler and Thorpe as the secret sonograph was first put to use.

You can see why this song was a fine subject for scientific analysis. It is simple enough to reveal structure and variation, but musical enough to be far from a call. Chaffinches do well in captivity, and through years of study and comparision between caged and wild birds, Thorpe gathered enough data to confirm what bird fanciers had known for centuries: a bird raised alone in a cage, with no contact with its own kind, will usually learn only a partial or incomplete song. In the case of the chaffinch, birds left alone will get the beginning right, but never quite manage that penultimate *che-wee-ooo*. Thorpe's research laid the foundation for what has since been observed and documented in most bird species: song is a mix of innate and learned parts, in a proportion that varies tremendously from one species to the next.

In their first few months of life, young male birds are much looser and more experimental in their singing. This is the sensitive learning period, when new sounds can still be learned. The sounds that come out were once called whisper song, and Thorpe christened it *subsong* to suggest its role as a stage on life's way to learning the song that matters. Ironically, subsong in the chaffinch can be much more extensive and complex than the full adult song, but it is not particularly noticed by other birds. It is clearly heard as a rehearsal, not a performance. When they are young, before they turn three months old, you can teach chaffinches to imitate other birds' songs, but soon they lose interest, because these alien songs will not serve them well later on.

What exactly is the service chaffinches need from their songs? Unlike calls, songs do not have a specific simple meaning that each syllable can reveal. Their basic purpose has been known for centuries: to attract mates, and to stand one's ground. Simple enough. Science has now worked on the chaffinch for more than forty years. Sonograms have been used to analyze the song, and tape recorders have been used to play songs back to birds, both male and female, to investigate how they respond. Their own songs, stranger songs, altered songs, typical songs, unusual songs, all have been tested relentlessly in what the field calls *playback experiments*.

Here is what recent playback experiments by Albertine Leitão and Katharina Riebel have revealed about the chaffinch: phrases 2 and 3 seem to be particularly important, what they call the "trill" and the "flourish." Thorpe knew that the *complete* song is necessary to serve both functions of territory defense and sexual attraction, because he had tried playing back abridged songs, and the birds didn't care much for them. But the new study shows that female chaffinches seem to prefer songs with a longer concluding flourish (phrase 3), while males react most strongly to songs with more trill (phrase 2) and less flourish. Yet if no flourish is heard at all, the males are not much impressed—perhaps it sounds too much like immature subsong. The title of Leitão and Riebel's article states their hypothesis—"Are good ornaments bad armaments?" This is some of the first research ever to show that male and female birds perceive the same song a bit differently. Or is it "quite" differently?

Bird song has long been thought to have these two imporant sometimes simultaneous functions. But this is a great puzzle: If song is sung both to attract mates and defend territories, why are there relatively few cases of specifically different songs evolved for each purpose? Why do most birds make do with one song for both tasks? And why do some birds, like the chaffinch, make do with simple songs, while others, like the nightingale and mockingbird, sing hundreds of variations all through the night?

Even the most voluminous surveys of the contemporary field of bird song research admit that no simple answer is forthcoming. Catchpole and Slater's *Bird Song: Themes and Variations* (1983) concludes that it may be "naïve to attempt a general, overall explanation for the evolution of song diversity . . . . We are left to puzzle over the resulting richness and variety that evolution has created." Keep in mind that this is from a science book! The finest compendium of modern bird song science, *Nature's Music*, edited by Peter Marler and Hans Slabbekoorn and published in 2004, continually reminds us how little we actually know.

Thorpe himself, a clear believer in objectivity and the quest for purpose, also believed in the inherent musicality of bird songs. After the chaffinch he embarked on an extensive study of the male–female duet songs of the African bou bou shrike. Thorpe was fascinated by the tonal quality of the duets performed by male and female birds, following regular patterns with very little variation. For this bird, he decided, musical notation was better than the sonogram in depicting what went on. Here are five different duets he heard, with four out of five transcribed as simple two-part antiphonal patterns:

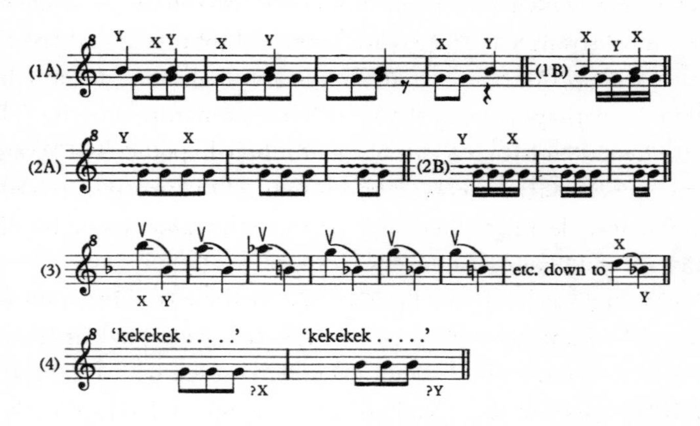

VARIOUS DUETS OF THE MALE AND FEMALE BOU BOU SHRIKES

These duets serve a third function, after mating and fighting: to strengthen the pair bond and help the two stay in close communication through dense thickets, where it is often hard for one bird to see the other. The most common intervals are the major third, as in 1a, and other consonant intervals as in 3. "Our judgment that bird songs represent music is not mistaken. . . . here we have art at least in its embryonic form." There is rarely an extreme dissonance. "To the musical listener these songs may seem over-harmonious; nevertheless it is the kind of harmony to which man aspired and which probably reached its peak in Mozart." The only fault in these songs is their brevity!

Now is this science or musicology? Marler saw two sides to his teacher's interests:

> You see Thorpe was an artist at dividing his life into separate domains, and I suspect that's what he was doing then. He saw an aesthetic dimension, which is fascinating, but then he leaves it. You spend years

> studying duetting in these shrikes and where does it lead you? My bias has always been to favor themes where you could see future prospects, the illumination of other domains, which leads one to sidestep extreme peculiarities which are sort of a dead end in a scientific sense.

It may not have led to a stream of eager disciples, but I would argue this is some of Thorpe's most important overlooked work. Is there no science in bizarre duets? There are fascinating questions here: Why would a species evolve a refrain that approaches the building blocks of our classical music? Did the basic physical laws of sound lead to simple harmonious beauty? Natural and sexual selection show the power of preference in the evolved world. A series of choices presumably leads to established variety. Yet we cannot be certain this is how this song arrived.

Since they are one of the more remote bird species, it is not so surprising that little more has been discovered about bou bou shrikes in the forty years since Thorpe made his report. Was it really a dead end? Yes—until 2004. That year, two German scientists, Ulmar Grafe and Johannes Bitz, were studying chorusing frogs in the Ivory Coast when they started to notice that every time they packed up their gear and left their research sites, a pair of bou bou shrikes began to chortle a particular song that they had never noticed before.

It seems that this trilling duet is an anthem of triumph, sung by the pair whenever they have won a conflict. This is the first animal species ever discovered to sing a specific duet only to announce victory over intruders, in this case: scientists. The victory song tends to carry further through the forest than the bou bou's other duet songs. It is usually sung from higher perches, to make it resound farther across the territories, letting all nearby bou bous know who's boss. Here is one bird song that has a more specialized kind of territorial function, a bit more like a call.

Is it possible to bridge the empirical and the aesthetic to find some reason for beauty? Communication across the greatest possible distance has been proposed as one of the reasons bird song is piercing and musical. Some studies show that the more a bird needs to defend its territory and proclaim itself across the greatest distance, the more complex and splendid its song. At Thorpe's laboratory in the 1960s there was a woman, Joan Hall-Craggs, who specifically focused on the musical quality of bird songs. She noted that the same characteristics that distinguish human song from speech also separate full bird song from subsong: The pitch is raised to those frequencies that are easier to hear. There is greater focus on specific notes, with sounds closer to flutelike tones. Song motifs become either longer or shorter, either a few pure tones or regular beats. Patterns arise in time and timbre. Put these two together and you get songs organized in phrases and forms. Only when a distinct form is established can specific songs be repeated and recognized. The more symmetry a song has, the more predictable it becomes.

The ubiquitous European blackbird's song has both melodic and noisy elements. Hall-Craggs concentrated on the melodic ones, using her expertise as a professional musician to try to explain why the song sounds musical to our ears. Previous studies had shown that blackbirds were taught phrases by their fathers and other neighboring adult males. This kind of learning appears in most bird species in only their earliest months. After the first season, the song is fixed. Not so the blackbirds. They learn new songs each breeding season—their sensitive learning period is seasonal throughout their lives.

Hall-Craggs listened for many years to just one blackbird, a wild one outside her home who was accustomed enough to her presence not to mind her listening and recording. This bird began one March with twenty-six basic motifs, mostly melodic bits of song. Some of these phrases, like the chaffinch song, had a single flourish at the end, a flourish not of a single note but a rasp or a squawk or something that sounded less musical to Hall-Craggs's ears.

At first the bird repeated each basic phrase a few times in succession, as if he were working his song up to snuff. This stopped after a few weeks. Then he worked at building a longer, more complex series of phrases, sometimes modifying the original phrases to create a more extended form, which peaked at the beginning of June. The noisy flourishes were tried and tested until just one version was decided upon. The song became progressively more organized, but did not reach any fixed form like that of the lyrebird, instead reaching a flexible form with a more well-balanced sense of melody than it had in the first weeks of early spring.

The following is an illustration of part of the process of song refinement, using both musical and sonogram notation. The blackbird would never reverse the order, singing phrase 12 before phrase 7. He always went from 7 to 12, producing the musical combination that is also more satisfying to human ears. Hall-Craggs noted that when the blackbird was responding to an intruder, his song remained loose, disjointed, staccato, sounding more harsh and primitive. The most musical renditions came late in the season—*after* mating was finished, at a time there was less need to defend the territory. Why did the song continue to develop after its presumed functions were over? Hall-Craggs did not attempt an answer, claiming to be a musician, not a trained scientist, with a sample size of only one bird.

Marler remembers her well. "Joan suffered in a way because she was not entirely respected. She was not a scientist. She did not have all that much data, yet she made very rich generalizations from the limited material that she did have." This may be why her work is rarely cited in the literature, even though she was one of the first to listen closely enough to a bird's song to decipher its structure.

If you are looking for music—not meaning—you don't have to back up your intuitions with reams of data. It may be enough to simply reveal them. Hall-Craggs considered the more musical song to be the more developed, and found this development increasing throughout the duration of each season. Some might consider her analysis too anthropomorphic, but she aimed for description, not explanation. She listened and took down the song, and found it to be far more musical than was necessary.

As a conclusion, she compared the form of that two-part blackbird phrase with two pieces of music, an excerpt from Bach and the well-

HALL-CRAGGS'S COMPARISONS OF
TWO BLACKBIRD PHRASES

known shanty "Blow the Man Down." Both have a formal rise and fall, a sense of beginning, end, shape and balance. The blackbird compresses these qualities into its own rapid sense of time:

The bird sang these same two phrases eighty-three times and never once reversed the order. Something about that sequence was more satisfying to him, as the A-B structure of those songs is to us.

What have scientists heard in blackbird songs? A more recent and equally assiduous listener to blackbirds, Danish biologist Torben Dabelsteen, did not consider the coda of this species' song to be merely a noisy flourish. Using more modern techniques of computer analysis, he shied away from such human categories of aesthetics as sea shanties. He identified three kinds of full blackbird song beginning with the more musical motifs, and ending with what he calls the "twitter," and two types of more limited song, one with just musical motifs, and another with only high twitters. Below are examples of high-intensity full song and twitter song. Dabelsteen was interested in what specific information the various song types might convey, and he claimed that they signify quite a lot: the singer's location, species, sex, territory, and degree of arousal, and the chance that sexual or aggressive behavior might be imminent. He found that the quiet twittering song communicates the highest arousal, and predicts the most specific range of aggressive responses: male versus male.

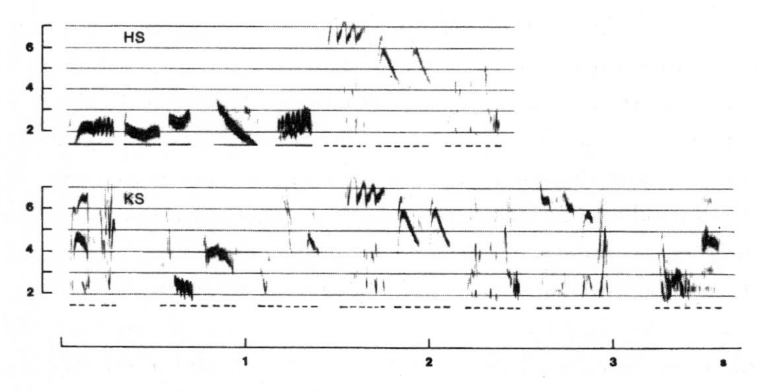

DABELSTEEN'S COMPARISON OF FULL SONG AND TWITTER SONG IN THE BLACKBIRD

The most belligerent behavior did not make use of the most extensive song, but instead something more threatening and noisy. Surprise? Not really. It's the same thing Hall-Craggs heard, although she was less interested in those "nonmusical trills." Dabelsteen also points

out that the gruff sounds can be heard effectively only at close range. The more melodic stuff carries much farther through the thickets, something not necessary for a territorial dispute but more useful for calling females near.

So what's the purpose of the more musical performance? Dabelsteen says long-range communication. Hall-Craggs supposes the bird is simply taking the phrases as musically far as it can—singing because he can sing. In a way these answers are complementary, showing two kinds of human perceivers—one with music in her ears, and the other with a theory of communication to test out.

Scientists who say they are investigating what actually occurs in nature caution that musicians and poets tend to hear what they want to hear, to extract some human meaning out of the world's alien inscrutability. Musicians remain enthralled by what seems unassailably beautiful about the sounds of birds, whether akin to noise music or dulcet melodies. But the scientist wants to know what the bird is listening for—is this a harder or easier question?

It may be impossible to escape the human perspective. For years scientists recording bird sounds tried to get as close as possible to the bird, to get the cleanest sound, with as little extraneous noise as possible. This led to a better recording, and a well-defined sonogram much easier to read and decipher. But do birds respond best to the crispest version of their own song? With blackbirds, Dabelsteen showed that when birds are at close range, they sing hissy and noisy sounds, acting out with sound to substitute for actual fight. The melodious sounds, in contrast, seem designed to be heard from far away. So wait a minute: If birds listen to their most stirring songs from greater distances, are they not hearing something very different from the clear, close-range song that scientists have been recording and playing back?

This was a revelation in bird song science, first pointed out by biologist Eugene Morton in the 1970s. He called this the *ranging hypothesis,* suggesting that one very specific way bird song works, to fulfill those basic dual functions of territory defense and mate attraction, is that the bird who is listening can gauge the distance of the singer from the amount of degradation in the song. Or if the listener is a male, he can tell how close his enemy might be. Songs that degrade faster through the bird's actual environment would be more useful for gauging distance. Those that carry farther would be more useful for functions that don't need to comment on distance.

The ranging hypothesis was first tested on the Carolina wren, a tiny bird with an unusually loud, warbling *teakettle teakettle teakettle* song. If you play a male Carolina wren his own song, he offers one of two possible responses. He might drop whatever he is doing and immediately attack the source of the song. Or he might stay put, cock his head, sing back in return, and then go back to his hunt for food. What determines the difference? Not the volume of the song, concluded Douglas Richards. Nor the length or complexity of what is sung. What matters most is how pure the song is.

A clean, close recording produced the most violent response, while a distorted, fuzzy one—the kind of sound you would hear if the other wren was far away in the thickets—produced a more gentle, song-matching response. Not that the responding bird himself sang a degraded song—he sang a clear song. But the other male, who started the banter, would hear it as a degraded song if he were far away. Thus the ranging hypothesis lends support for the function of bird song as territorial sonic contest between rival males.

Some scientists believe gauging the distance of the singer matters much more than exactly what is sung. Birds who sing songs that carry long and far through the trees may be doing it as part of a sonic "arms race," using a musical weapon that is hard to pinpoint. Predators (or bird watchers!) might not be able to tell where the song comes from, but possible mates would know their bird is within range. A bird with a vast repertoire of songs might confuse listeners with its maze of tunes, some familiar, others not. Their various airs might baffle listeners, tripping up whatever message should be conveyed. Between male and male, there are more cases of song-matching and exchanging. Female-male duets tend to be more fixed, enhancing that pair bond like the harmonious tunes of the bou bous.

The standard degradation of usual songs enables the listener to tell how far away the singer is, leading to a vicious or gentle response. The more resounding songs evolve as a kind of avian deception. Sing your song in a way that it degrades as little as possible, or vary the song enough from what is expected, and you can gain the edge over a rival without getting too close.

If music be the food of war, sing on? This is no soulful view of what makes these special songs beautiful. Functional explanation draws us away from the song itself and focuses instead on what it is meant to accomplish. The ranging hypothesis is one of those intriguing, counterintuitive explanations that pushes science forward. It matters not how beautiful a song is, but how identifiable it is from very far away, when it is only a faint echo of its possible self. What if I'm close enough to hear the full bird loud and clear? He'd rather tear my ears off.

The ethereal swirls of American forest birds—the wood thrush, veery, and especially the hermit thrush—are most interesting to consider in light of the ranging hypothesis. Why have such odd whortling tunes inspired the most moving words from naturalists and poets and some of the most intense calculations from scientists? I will travel through a history of responses to these mystic sounds before they are played into the sonograph.

Thoreau wrote of the song of the wood thrush, "It lifts and exhilarates me. . . . It is a medicative draught to my soul. It changes all hours to an eternal morning. It banishes all trivialness." It happens so fast that several notes seem to be produced at once, like otherworldly chords, and the whistles through the syrinx resemble experimental flute techniques championed by today's most radical virtuosos. John Burroughs was bewitched by the deep simplicity of the hermit's sound. "'O spheral, spheral!' he seems to say; 'Oh holy, holy! O clear away, clear away! O clear up, clear up!' interspersed with the finest trills and the most delicate prelude." Burroughs did not find the song full of any passion or emotion, but that may be what makes it more musical than any other American bird song: It is far from the emphatic directness of a bird call; in its precise but enigmatic form it seems to be about something entirely different, a "calm, sweet solemnity" you reach in your finest moments. Beside it, thought Burroughs, listening to one on a cool, full-moon night, the pride and bloom of civilization seemed trifling and cheap.

The hermit is the second bird whose spirit and rhythms Walt Whitman learned from, in his poem "When Lilacs Last in the Dooryard Bloom'd." The story is by no means as direct as the boy mesmerized by the mockingbird. In this case, Whitman was looking for a bird as an image of something ultimate, and John Burroughs suggested to Whitman himself that the hermit thrush was the bird to make use of:

> Loud and strong kept up the gray-brown bird,
> With pure deliberate notes spreading filling the night. . . .
>
> As low and wailing, yet clear the notes, rising and falling, flooding
>   the night,
> Sadly sinking and fainting, as warming and warming, and yet again
>   bursting with joy,
> Covering the earth and filling the spread of heaven. . . .
>
> Sing on dearest brother, warble your reedy song,
> Loud human song, with voice of utmost woe.

> O liquid and free and tender!
> O wild and loose to my soul—O wondrous singer!

For Whitman the song is strangely human, though I maintain it is powerful because it is an alien music: free, wild, loose, and wondrous.

Canadian naturalist M. Chamberlain noted this in 1888:

> The music of the Hermit never startles you; it is in such perfect harmony with the surroundings it is often passed by unnoticed, but it steals upon the sense of an appreciative listener like the quiet beauty of a sunset. . . . On one occasion an Indian hunter after listening to one of these choruses for a time said to me, "That makes me feel queer." It was no slight influence moved this red-skinned stoic of the forest to such a speech.

It is a querulous, queasy sound, which carries far through the distant leaves in a slinky, slipped-into-the-trees kind of way.

We have already seen how the hermit carried Schuyler Mathews into poetry. When Aretas Saunders gets to the hermit thrush, he does not seem particularly impressed. It's just another warble for him, not the paragon of avian music. Maybe this is because of the notation he was wrapped up in, revealing just a bunch of squiggly lines rising up and around, not the glimmers of a pentatonic scale so familiar from the world of human music that warbles beyond the timbre of any simplified flute. With lines and words we still get a sense of the chortle:

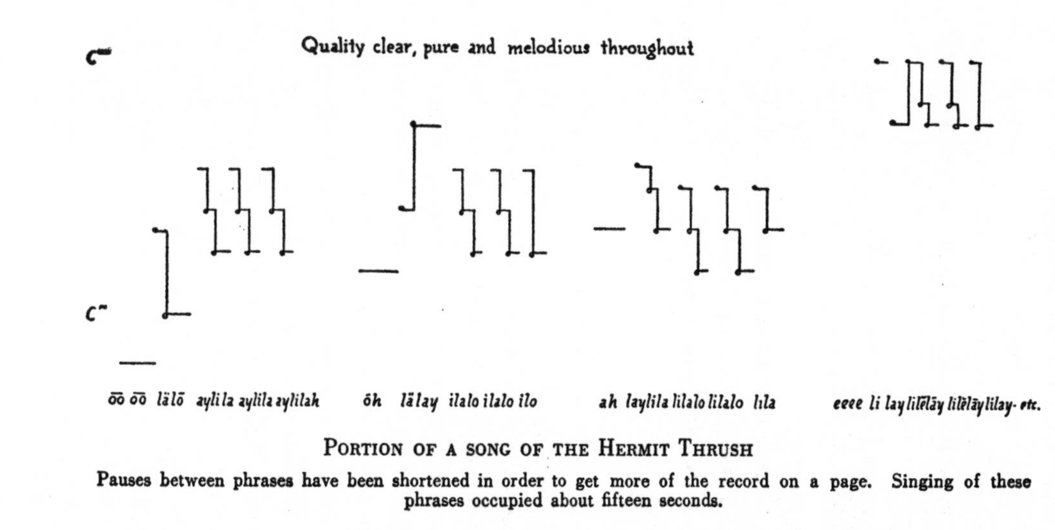

PORTION OF A SONG OF THE HERMIT THRUSH

Pauses between phrases have been shortened in order to get more of the record on a page. Singing of these phrases occupied about fifteen seconds.

Another made-up language emerges, of *lälay ilalo ilalo ilo*—more syllables straight out of Schwitters. Even the more staid T. S. Eliot heard the primacy of elemental nature in the hermit thrush's peculiar tune:

If there were water
And no rock
If there were rock
And also water
And water
A spring
A pool among the rock
If there were the sound of water only
Not the cicada
And dry grass singing
But sound of water over a rock
Where the hermit-thrush sings in the pine trees
Drip drop drip drop drop drop drop
But there is no water

Not a bird he heard in England, that's for sure—must have been a lingering memory from his homeland. Eliot makes the brown bird large, with a desire to grasp all of nature in this elusive music. We hope the watery sound contains all.

When all these human observers praise the music of the hermit thrush, they are most impressed by the fact that it sounds musical on its own terms, not according to the rules of human music. This bird seems to be forming musical phrases that dizzily swirl with an internal logic and structure far removed from our own ideas of what music must be. We admire its composition because it is a bird's music, not a man's. That's what made the native hunter so uneasy.

The Hungarian musicologist Peter Szöke achieved the most detailed combination of the sonogram with notation when he analyzed the musical depth of the hermit's song. Why did he sit in Budapest working on this distant American song? He was impressed by the specific musicality, the order and form not quite able to be grasped by the naked ear. Szöke was able to adapt a tape recorder to slow down the sound while adjusting the pitch to a range that best suited the human ear. This is something any computer with the right software can do today with ease, but in the sixties it required some jerryrigging. Once he had stretched out the song to thirty-two times its actual length, one and a half seconds becomes nearly fifty seconds. The result is an enhanced revelation of pitch, structure, and ornamentation of the most precise subtlety. Szöke's version, with its long declamation and cool ornamental bends, resembles a shepherd's lilting tune from one of Bela Bartok's collections of Hungarian folk music. He focuses on the rhythm of the song—not the notes the bird sings—which Mathews simplified into a pentatonic scale like those blackbird-Bach-sea shanty melodies. Recent studies show that the hermit thrush does divide its tunes into five-note scales, but not quite the tempered black keys on the piano. Instead the pitches are closer to the pure partials of natural harmonics, like overblowing on a flute or tinging a guitar string. Szöke transcribed a musically rich statement of deep rhythmic complexity, with many of the accoutrements of the modern classical music of his time. He was most impressed with the following slowed-down, transcribed version of the hermit thrush song. "This is the highest summit in the evolution of animal music so far known to us. . . . It is true micromusic."

Do all these fleeting variations matter to the bird? Can she actually hear very quick sounds with any more detail than we can? That is the kind of question a scientist could try to answer using carefully designed playback experiments. Humans can hear frequencies from just a few cycles per second (Hertz), the lowest bass notes, to 15000 Hz, tiny high whistles far beyond the reach of a piano or flute but included in their overtones. Most birds only hear from 500 to 6000 Hz, but within that range they are much more sensitive to variations in sound than we are. Jeff Cynx at Vassar College has demonstrated that birds do not respond well to transposed versions of their own songs:

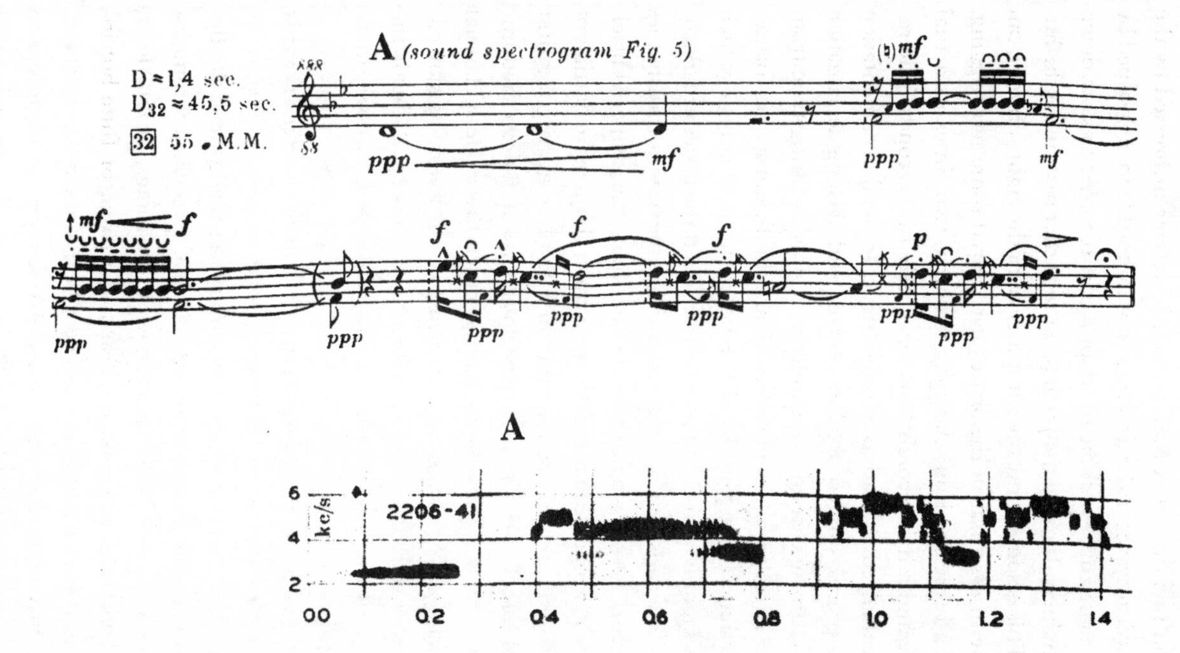

PETER SZÖKE'S NOTATION OF A SLOWED-DOWN SONG OF THE HERMIT THRUSH, COMPARED WITH A SONOGRAM

they need absolute, not relative pitch to know what they're hearing. They cannot hear intervals as such so well, but Robert Dooling has shown that they are much better than humans at discriminating between the kind of noisy sounds that give complex sonograms in fractions of a second. Birds can discriminate time differences between one and two milliseconds, while humans can only do three to four milliseconds. They're twice as good as we are.

So although hermits might not recognize *all* the musical details that Szöke has elucidated, they are able to hear far more detail in their original, real-time song than we are. Szöke found all that complexity in one single hermit thrush phrase. How might many phrases fit together? Is there a grand order to the bird's singing of its twenty to thirty different song types?

Charles Dobson and Robert Lemon analyzed the series of songs of several of the most interesting-sounding American thrushes: the wood thrush, the hermit, the veery, and the robin, using a simple mathematical model called the Markov chain. It's similar to models that try to predict tomorrow's weather based on today's. First identify a set of possible choices—clear, partly cloudy, rain, or snow. Then collect data on past weather fluctuations, as much as you can get. Then see how many days previous to the current day you need to make an accurate guess of today's. If all you need is one day back, then it is a first-order Markov chain. If you need two days back, then it's second order. Simple mathematics helps to show that apparent complexity in choice of weather (or song) might require a much simpler organizational principle than one would imagine.

The bird knows, say, twenty short phrases. It sings phrase 4. How does it know what phrase to sing next? If the rule governing the process is a first-order Markov chain, then two things are true about it: (1) No memory of previous phrases sung is needed. The next phrase sung depends purely on the identity of the current phrase. If you need to know which one or two phrases are sung just before, then it becomes a second or third order Markov chain. (2) The rules don't change as the song goes along.

Dobson and Lemon found some level of Markov order in the songs of all the thrushes they studied, with species with simpler songs having simpler levels of chains, and the hermit thrush having the highest level of variability. Similar Markov modeling has enabled computers to distinguish between the music of Haydn and Mozart, and also enabled machines to create new compositions that resemble the arrangements of phrases that various composers have used. It is not all that surprising that this kind of order should be found in the sequences of bird songs, reinforcing our idea that there is some music in the overall scheme of them after all.

Are such simple models really the key to complex-sounding bird songs? Mathematicians like Markov chains because they get fast results. But they are not particularly deep. With a Markov chain model, a computer can generate nonsense poetry with little effort, producing something that superficially resembles human language. But we can quickly tell it doesn't make sense. When the computer emulates a sequence of bird songs, it is harder for us to tell. It gets easy to become impressed with our own modeling prowess, but birds are rarely fooled.

Consider another thrush song, the swirling descending phrase of the veery. It's even more queer and querulous than the hermit thrush. Mathews drew the song thus:

And Saunders so:

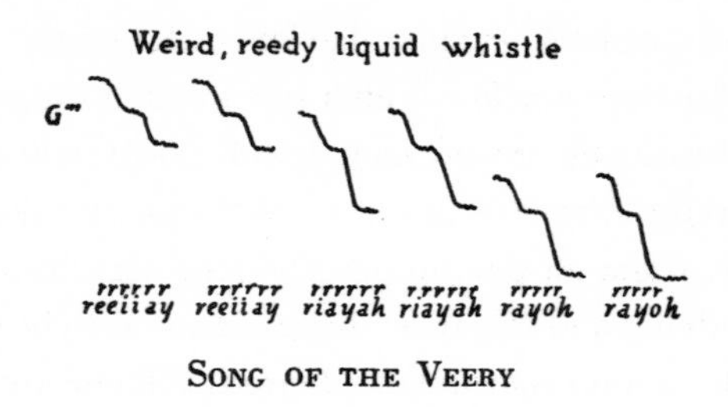

SONG OF THE VEERY

. . . the song like a sweeping round sigh descending through the trees. Not as musical to our classical aesthetes but more like a wash of synthesized atmosphere present in the electronic music of today. I slowed it down and heard a lilting line like a phrase out of Miles Davis's electric fusion period:

I really didn't expect something like that to come out—a swinging melody that changes from C minor to G7 midcourse! The rhythm is also a surprise. Szöke had Bartok on his mind; maybe I'm so steeped in jazz that I hear syncopation in every bird. I doubt it, though. This veery does stand out.

Lemon wanted to see how the veery reacted to tripped-up versions of its own song, played back through the forest. Shorter, longer, higher, lower, faster, slower, all mixed up—he tried them all. The experiment concluded that the birds were most responsive to the playback of edited songs that maintained the order of the syllables: the whole *shape* of the phrase. The higher pitch was more important than the lower. Take out the lower pitch and the birds still responded, if the rest followed the correct structure. What about the most characteristic quality of the song to human ears, the rapid modulation that makes it so eddying and ethereal? The birds need to hear this as well to know that the sound is meant for their ears. These are all aspects of the sound that can be heard, even in a degraded form, far away through the trees. And they sound like a veery, and no other thrush. *Reeiiay rayoh!* She knows veery music when she hears it. The sounds that are meant for her have a specific shape and form. Is there any significance in the actual quality of that shape?

In one of the last papers he published, W. H. Thorpe collaborated with Joan Hall-Craggs once more and returned to the general topic of pattern perception. They were inspired by the view of Gestalt psychologists that a tune or melody is always more than the sum of its parts, that it must have a particular feature that defines and identifies it, some gestalt that is difficult to extract or describe. Playback experiments have convincingly demonstrated that birds of many species can hear what's right for them, and what's wrong. We notice the spiraling swirls of the veery, we know it's a veery. The veery adds a deeper knowledge: it hears the same song but knows it is right and necessary, the only song worth singing and hearing.

Thorpe and Hall-Craggs conclude with a return to her famous blackbird. Certain blackbird sounds are introductory themes in the song, others always appear as a conclusion. They consider this blackbird's song as a gestalt, which only makes sense when whole and correct. But the bird sometimes made mistakes. Often when this happened he started again, just like a human musician, repeating the phrase until he got it right. Once after not singing for a few days the bird lost the ability to complete a particular phrase, and whenever he got to the difficult part, he started to falter, gurgle, and repeat a few notes. In the end he was unable to sing the song as perfectly as he could before. Eventually he reached a compromise, changing the song into something easier to sing. But he couldn't just replace the difficult syllable, he had to eliminate one whole phrase. Why? Because there was a right way to sing the phrase, and when he could no longer quite do it, the whole section had to be cut. The bird, they wrote, "was literally out of practice."

The closer we listen to the intricacy of a bird's song, the more we find evidence that there is a right and wrong form to the sound, just as in a piece of human music. The technology of recording made such close study possible, and the visual representation of the sonogram made things we humans are unable to hear suddenly visible on the page. As we learn to read them, these diagrams show patterns perceivable by us, and patterns perceivable by the bird. These two are not always the same.

Peter Galison has pointed out that objectivity—a sacred goal of science—is in fact a historical value that has developed over time together with scientific judgment. Scientists train themselves to interpret data, which is often displayed in novel visual ways like the sonogram. It is no easy impartial process to interpret what a sonogram reveals. Peter Marler recognizes that it is a constructive act. "You choose a bandwidth for the analysis, and having made that choice, you bias the way in which the information will be portrayed." As the tool was developed and refined, so was the ability of scientists to discriminate what they could see in the image produced.

Galison likens this process to art criticism or aesthetic education. The scientist doesn't just strive to discover what is *there,* in the supposedly unassailable objective picture of some fraction of the natural world. The sonogram must be adjusted, drawn upon or filtered to produce an image that represents a conception of what a bird is doing. If the technique works, and we learn how to read it, then it is more standard than any individual's mnemonic phrases or musical transcriptions. But it is still a picture, and for a picture to become a language, somebody must discover the rules. We will never become birds by picking apart their songs.

Here's how the lyrical French historian Jules Michelet described, around the same time as Darwin, the intensity of the avian aesthetic—"Winged voices, voices of fire, angel voices, emanations of an intense life far superior to ours, of a fugitive and mobile existence, which inspires the traveler doomed to a well-beaten track with the serenest thoughts and the brightest dreams of liberty!" Listen to birds, think of them like that, and you too will want to copy them and join in their music. Is that not what they do themselves?

THE MARSH WARBLER

CHAPTER 5

Your Tune or Mine?

NO ONE FUNCTION CAN EXPLAIN the musicality of the avian world. Nowhere is this more apparent than among those birds who construct their songs from pieces of the songs of all the other birds they hear. Consider the European marsh warbler, *Acrocephalus palustris,* which breeds in the wet meadows and thickets of Northern Europe. One by one it repeats nearly all the sounds of all the other bird species that live in its habitats, one after another with little recognizable pattern or repetition, packing a few tiny fragments into every second like a bird song identification tape played at double speed. This sonic acrobat sings on and on, a flurry of stolen licks going by so fast it is a feat to pick them out. He's arrived in Europe after a winter in East Africa, and he perches in a bush, streaming out his rapid outburst. This excerpt, all in the matter of a few seconds, begins with two alarm calls of the greater whitethroat, followed by the rapid four-part (123412341234) social song of the barn swallow and then a few alarms of the great tit (a European chickadee):

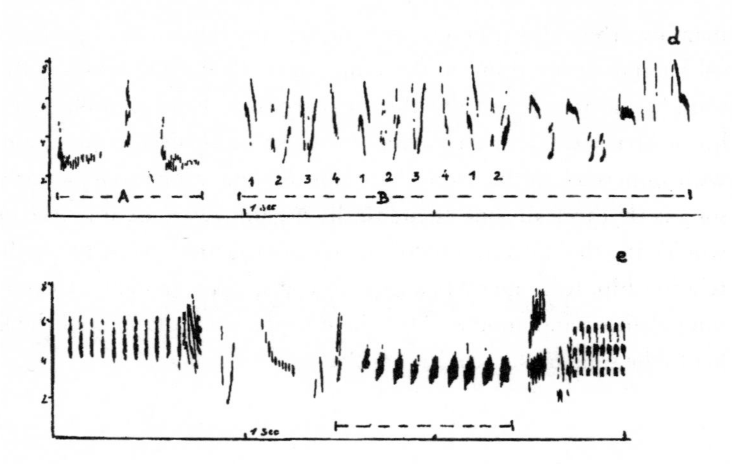

About half of the marsh warbler's song is a rapid, incessant recall of every other bird sound around. For decades it was thought that the rest of the song—an equally confusing melange of cheeps, trills, slides, and squawks—was the authentic marsh warbler tune because it did not sound like any other native birds. It turns out that is not the complete story.

Françoise Dowsett-Lemaire first heard the marsh warbler as a fifteen-year-old girl outside her childhood home near Liège, Belgium. I reached her by phone in the south of France, and asked how she was able to grasp so complex a song with her ears alone:

> I was impressed by its imitative ability and wrote down in my notebook a longish series of European bird calls I could recognize in its song. Most imitations are extremely brief (less than one second), but I think my ears were especially trained in those days to pick up motifs so short and fast. Thirty-five years later I find it much harder to identify these brief motifs. My hearing isn't what it used to be.

This little brown bird of dark, soggy thickets became the subject of Lemaire's doctoral thesis at the University of Liège. With her keen ear and the help of a spectrograph, she was able to ascertain that the marsh warbler ably imitated most of the bird calls and songs of its local habitat—everything in the range from 1500–8000 Hz—all those sounds that it is physically able to reproduce. European blackbirds, house sparrows, and tree sparrows were most often copied, along with linnets, skylarks, stonechats, and magpies. The marsh warbler song is rhythmically and thematically organized. Sometimes the bird would sing the call and song of one species together, recognizing their relationship. In some cases a smooth transition is made between one song element and another. Here he morphs from the song of the linnet to the song of the willow warbler:

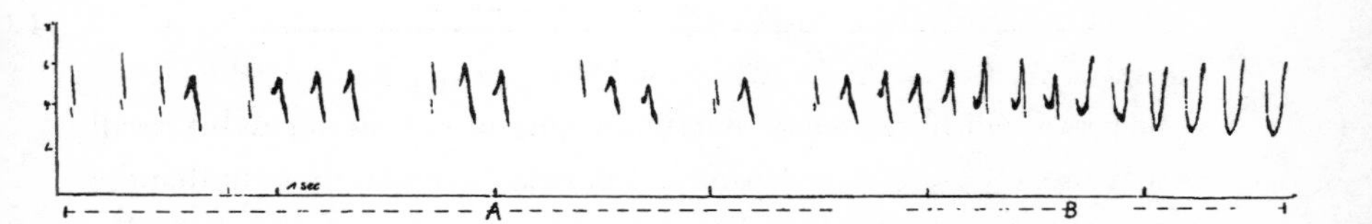

What about the other half of the marsh warbler's song? Lemaire suspected that the rest might contain imitations of African bird species that were picked up on the warbler's winter migration sojourn. Only problem was, no imitating bird had ever been known to do such a thing. Most expert mimics either do not migrate or learn most of their material in their first season, before they take off for the tropics.

In 1976 Lemaire took her first trip to East Africa, the winter grounds of the marsh warbler, where she learned the sounds of common birds in the area. When she returned to Belgium the following year, Lemaire was astounded to hear the warblers perfectly imitate many of these African birds. "Calls of resident European birds and of African birds are evidently memorized at different times of the year and their assemblage into motifs take[s] place later, during the transition from 'mixed' to adult song." Woven into the marsh warbler's extremely complex song she heard black-eyed bulbul notes and bleating bush warbler calls, along with such colorfully named singers as the blue-cheeked bee-eater and the fork-tailed drongo. The more common and noisy the birds in the warbler habitat, the more likely they are to be imitated. The marsh warbler turns out to have no original song syllables at all. It builds its whole repertoire out of others' material! The originality lies in the expert mimicry and virtuosic recombination of all the sounds in its vocal range. It even imitates birds it can only hear en route to its winter grounds: passing through North east Africa, it picks up tunes from the Boran cisticola and the vinaceous dove. Françoise Dowsett-Lemaire had discovered the one bird in the world who can recount its migratory path as a kind of songline, where the journey is mapped into the music itself.

Why has this bird evolved such an incredible mimetic song, perhaps the most complex in the world? Not for mating purposes. Female marsh warblers choose mates based on the size of the male's territory, not the quality of his song. The females seem rather uninterested in its awesome complexity. Says Lemaire, "The larger the territory, the more potential nest-sites it contains, and the more likely the first female who comes in will stay. Besides, as the song is so complicated and it takes over thirty minutes of continuous singing to get the full repertoire, females would need to sit and listen for ages to evaluate a male's musical skill. Of course they do no such thing." As soon as a female appears, the male stops his singing. Concert over. He then devotes himself to helping her find the best nest site, giving only brief snatches of song along the way. She may never get to hear what her mate can do.

Females, Lemaire notes, occasionally sing the same kinds of songs as males, but for no more than a minute at a time. She believes they have the same sound knowledge as their mates but have less need to use it. A similar pattern is found throughout the sexist world of bird songs. Females can sing if driven to do so, but most of the time they have too much else on their minds and not enough of the right hormones to pull them on toward music and away from practical life.

The song of the marsh warbler is far more complicated than it needs to be for the presumed purpose of defending territories and attracting mates. Lemaire also noticed groups of neighboring males joining together for "periods of peaceful singing when they are off-duty"—that is, not bringing food back to the babies or sitting on the nest while their mates take a break. Two to four birds will sit on branches close to one another and sing together. Only on clear sunny days, never on tough, rainy days, when all their energy is needed to find food. Lemaire doesn't believe these group choruses are song bouts or challenges, but rather a kind of social play. Male marsh warblers "enjoy singing and must realize in some way that music is fun. There is no doubt about that." No doubt, that is, until one tries to make that kind of statement in the context of science.

How could you prove that birds have fun while they sing? Measure some rise in their emotional states like the weight of those nightingales? Finding factors to evaluate is the essence of science, to turn feeling into rule, tendency into impulse. Peter Marler muses, "Birds are driven, that's something you can't deny about song; it is a behavior subject to *strong* internal motivation. Is that drive something like an emotion, with subjective connotations to it? We tend to assume a bird is being joyful. This may or may not be true, but that puzzle has always interested me." It is a game with many pieces missing: science has not evolved to the point where it is able to calculate joy.

For Dowsett-Lemaire, the marsh warbler's joy in singing is obvious. With her unusually discriminating ear, she listened deeply to this bird for nearly fifteen years, with a level of attentiveness unparalleled in recent bird song science. She found a song that is organized and creative, yet not easy to register as musical in human terms. It is a booming, buzzing confusion of snippets from the avian soundscape of two continents. A recording of the song successively slowed down—half speed, quarter speed, eighth speed—reveals further elements of complexity in an almost fractal manner. It is astonishing that this bird is able to assimilate so much of what it hears, and to tease with it at such a high level of facility.

That birds imitate the sounds beyond their species was one of the first aspects of bird songs noticed by human observers. In ancient

Rome, Pliny had a rather high-minded view of the intellectual angst suffered by magpies—"They get fond of uttering particular words, and not only learn them but love them and secretly ponder them with careful reflection, not concealing their engrossment. . . . It is an established fact that if the difficulty of a word beats them this causes their death!" The first truly scientific study to deal with bird songs, *The Evolution of Bird-Song* by Charles Witchell in 1896, devoted a considerable number of pages to mimicry. Witchell believed that many British birds imitated other birds in their habitat, and this was the first proof for him that birds learned rather than inherited their songs. He often heard mimicry where it wasn't actually there, but the idea that birds were attuned to their surroundings gave specific credence to the evolutionary dream—that animals evolved in response to their environment through time. "We need not wonder that the cries of birds are so often somewhat similar to the sounds which the birds themselves experience daily . . . , and that the voice of the bird has been thus attuned to harmony with neighbouring sounds, just as its colours so often blend with those of its surroundings."

Joining the new theory of natural selection to the attentiveness of a naturalist, Witchell was the first to want to be empirical with his subject, but like many a transfixed listener, he would sometimes wax poetical. He heard murmurs of water in the nightingale's tones, its rhythms echoing the shafts of moonlight tracked through midnight trees. "He is pleading now! but no, he is declamatory; now weird, now fierce; triumphant; half-merry: one seems to hear him chuckle, mock, and defy in almost the same breath."

Various adaptive explanations for mimicry have been proposed since, though none really explain why it is so prevalent. As territory defense, what could be better than scaring all kinds of species away by making a mockery of their styles? That is why mockingbirds are often considered so aggressive. They take so many snatches of so many different songs that it's hard to imagine any bird wanting to fight back, at least with music. Some have called this the "Beau Geste" hypothesis, after the famous film of the lone French legionnaire who defends a fort all by himself by imitating many soldiers' voices to convince the enemy that the fort is full of troops instead of just one man. A nice idea, but no one has been able to prove it works for even a single species of bird.

Mimicry might be most common among birds who need to communicate across long distances where it is hard for them to see one another. If so, it might work a bit like the distinctive *whoorbles* of those thrushes in the previous chapter, being the best kind of sound to carry far through the trees. This turns out not to be true either—mimics live in all kinds of vegetation, dense to thin. Mimicry might exist to distract predators or to lure in prey. Never proven. Birds more often imitate other birds' *calls* than their songs, which ought to be able to lead us to prove something. Calls are much more specific in meaning, but the meaning doesn't seem to be copied along with the sound. A parrot doesn't respond to a lyrebird's perfect copy of its call. Why aren't the species being imitated at all interested in the imitations? Clearly they know something we don't.

Out in the fields that one night, John Clare heard a nightingale who set him on a lifetime course of poetry. To enjoy the music of a bird, all you need to do is listen. But to initiate a science of bird song mimicry, you must be prepared to observe a bird's whole behavior over a long period of time. Much more is known about birds that do well in captivity than about the reticent ones who sing only in the wild. Some of the most intriguing studies of how mimicry is structured have been done with a bird that—at least in America—is one of the most despised of introduced species, *Sturnus vulgaris,* the European starling.

Edward Schieffelin was an eccentric nighteenth-century New Yorker with a curious mission: to introduce into the New World all the birds mentioned in Shakespeare. In *Henry IV,* Hotspur, forbidden by the king even to speak the name of his despised brother, Mortimer, comes up with an elegant solution—"I'll have a starling shall be taught to speak nothing but 'Mortimer,' and I'll give it to him to keep [the king's] anger still in motion." On the basis of that single reference, Schieffelin released between fifty and two hundred European starlings into Central Park. The starling gets its name from the tiny white stars that gleam, only in summer, against its glistening black feathers. They merit only one reference in Shakespeare, so clearly the Bard didn't think all that much about them—and Schieffelin did not foresee the consequences of his supposedly civilizing deed.

No one back then realized that the introduction of just one hardy, omnivorous species in a fresh environment might upset a delicate ecological balance, displacing scores of native songbirds and devastating acres of cropland. Central Park's literary flock spread across America. In five years they made it to Brooklyn, in twenty to Niagara Falls. After fifty years they were sighted in Colorado, and today there are two hundred million starlings in America, all the way from Florida to Alaska, making up one-third of the planet's population of this nearly ubiquitous bird. Mortimer indeed. The immigrant success of the starling and the English house sparrow help to explain why so many of those intriguing species in the field guide are harder to see and hear. Starlings are noisier, louder, and more social, and they travel in flocks a hundred thousand strong.

In America the song of the starling has not been much appreciated, perhaps because listeners often hear it as a rush of garbles and squawks coming from a tree filled with several hundred of them immersed in chatter. Less prejudiced Europeans have called starling song "a lively rambling melody of throaty warbling, chirruping, clicking and gurgling notes interspersed with musical whistles and pervaded by a peculiar creaking quality." The song mixes mimicry with its own unique tones and textures. Starlings eat everything, and they absorb all manner of peculiar sounds, choosing those that fit their own aesthetic.

Here is a song whose structure has been painstakingly decoded in recent years, in part because the bird is so adaptable and easy to observe. The most comprehensive deciphering of the structure of starling song has been carried out by another Belgian biologist, Marcel Eens. In 1997, in an eighty-page article, Eens concludes that a full star-

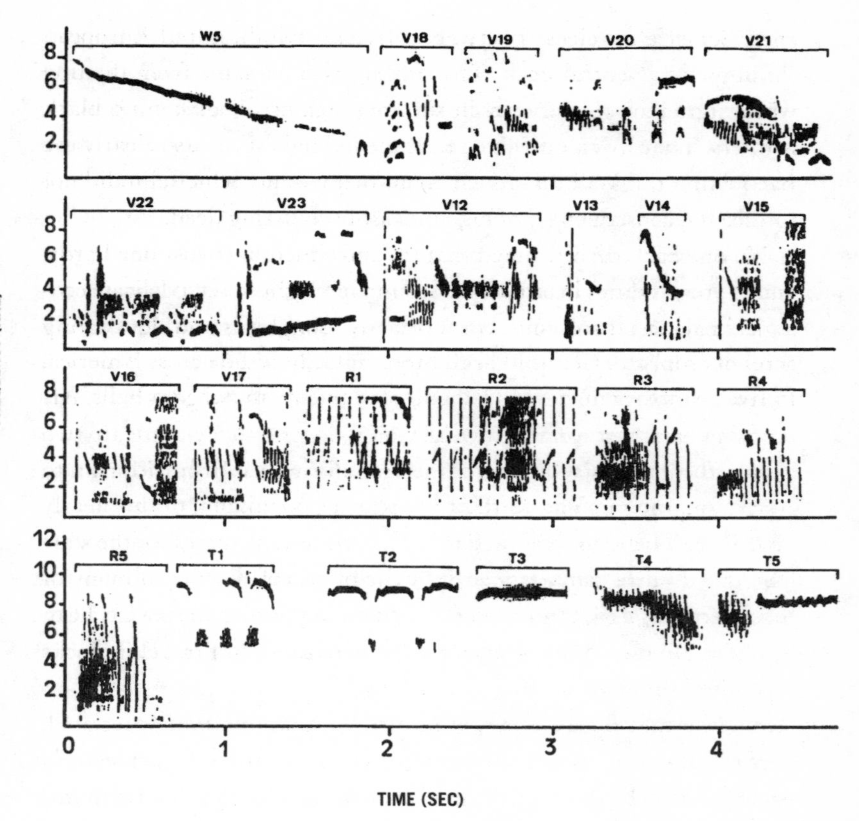

EENS'S SONOGRAM OF THE SEQUENCE OF PHRASE TYPES IN A STARLING SONG

ling song, which takes about a minute to sing, is composed of four distinct kinds of phrases, which are generally sung in the sequence shown above, although each type of song is repeated two or more times before the bird moves on to the next type: First, one or two descending whistles, out of a repertoire of two to twelve different kinds; then a quieter, continuous warbling, in which imitations of various birds living in the starling's territory are often inserted; the third part of the song is a series of rapid clicks, up to fifteen per second, a rattling or ratcheting with no clear breaks between; finally, the song concludes with loud, high-pitched squeals, repeated many times. This is the loudest part of the song, a strong, noisy but clear conclusion. Here is the whole thing: W = whistle, V = variable, R = rattle, and T = "terminal high-frequency song type." The repeats of each type are left out.

Now that you've got that, go outside and listen to a starling (wherever you are, one is likely nearby), and you will immediately hear things you did not hear before. There is a plan to what before seemed like noise. After the initial descent, the song gets faster, louder, and higher as it proceeds from beginning through middle to end. There is a clear dramatic form, with the exact content of each type of phrase varying greatly from bird to bird.

Further studies show that these songs have little influence on male territorial interactions, so their purpose is presumably to impress the females. The extreme differences between individual songs are thought to help females identify their own mates. This implies that each bird has a signature song that shares a parallel form with all the other starlings, while possessing a distinctive series of original phrases and imitations. In fact, a male starling's song in its most developed form is usually directed only at its mate, not at other females.

Starling couples are much more talkative than marsh warbler pairs. It is believed that the complex starling song advertises some part of a male's overall quality as a mate—but what? We have no sure idea why these songs need be so intricate. But there is no reason to shrug off that uncertainty with starlings. They are plentiful and easy to raise in captivity. Why not get to know them a bit more? Their relentness noise might get even more interesting.

Meredith West and Andrew King have raised and studied nine starlings for ten years at the University of Indiana. Four of the birds were kept to themselves while the other five lived in close proximity to their human caretakers, with extensive and friendly bird–human interaction. There were no specific attempts to teach the birds to sing or to speak particular words. Why not? West and King did not want to force-feed any sounds to their starlings. They let the birds share in their daily life without a detailed program in order to hear what they would pick up when left to their own devices. Only the five birds that had extensive daily contact with people learned to mimic human sounds. They would recognize simple phrases and recombine them in odd ways. "Basic research," one said. "Basic research, it's true, I guess that's right." One bird, which needed to have its claws treated for an infection, squirmed while held, screaming, "I have a question!"

Most of the time, though, their song followed the same form Eens described above: whistle, warble, clicks, and screech, but instead of just mimicking bird sounds in section two, they would mix in human phrases with odd, decisive endpoints. One bird often whistled the first notes, not the words, of "Way down upon the Swa-" without ever feeling inclined to add "nee River," even after hearing the whole phrase practiced hundreds of times on the piano. I think this shows a clear starling aesthetic: The bird just liked that shortened phrase better. "Way down upon the Swa. Way down upon the Swa." You try it. Has a nice ring to it. I can easily imagine it mixed into the dada of starling sounds. The starling doesn't practice to get the song right, like Hall-Craggs's blackbirds. He grabs what he wants and slots it right into that strange starling music, a style unto itself. Here's how West and King describe life with a starling:

> It was easy to forget the bird's scientific role as he sat with us while we had our morning coffee, took a shower with someone in the household, mimicked a child's crying, and attempted to join a lab meeting. . . . Starlings seem to see any setting as potential vocal turf. The most obvious sign that they are in a music-making mode was when they were quiet, cocking their head to and fro listening to whistles, music, maybe the teapot. . . . Indeed, the best way to quiet a noisy starling is to feed him a new sound; he or she must stop vocalizing to digest the vocal bite.

Once the bird decides to assimilate the new sound, he works it out of its human context and into the starling world. Many of his utterances come from a surreal place where bird and human collide. A sound like *breep, beezus, breep, beeten, beesix* might suggest the whole process of bird composition in action, then you might suddenly hear, "Do they have a toll-free number?" and you know just where that comes from but not why.

Starlings, like pet parrots, throw out a sound to see what happens. King and West call this *social sonar*. In addition to human words, their starlings also mimicked squeaking doors, clanking dishes, barking dogs, smacking lips, and human gulps. A student in King and West's lab, Marianne Engle, confirmed this hypothesis in the research done for her dissertation. She carried on her teachers' work and raised several starlings at home. Her starlings tended to imitate sounds directed at them or related to human interaction with the birds. To get more attention? To expect to hear human sounds back? Starlings in the wild, like many birds with complex songs, share certain elements of their songs with their neighbors. They live in large flocks, so the sharing of elements from song to song may indicate some sense of familiarity when the song is heard.

Eens did prove that the longer and more complex a starling's song, the more mating success he will have. But he did not study the actual content of that complexity. Why? Too many variables. As Peter Marler told me, "The starling's song is at the very limit of human comprehension." Even a song of one minute's duration taxes our analytical methods. Those who have lived with starlings for long periods have more than data on their birds. They have experience, and with experiences come stories. Marianne Engle noticed that the mimicked elements in the song are sometimes used separately. Here's how it worked with one starling named Elmer: "One of his favorites was a sequence of kisses, like one would use to call a cat. If Elmer was already singing, I found I could approach him and make the kissing sounds, and he would often immediately repeat the sound back to me, interrupting the rest of his song mid-stream."

Kissing seems to have some particular relevance to starlings; West and King noticed the same. Another bird would imitate the soft sound of the fluorescent light above his cage, especially one time when the power was out, as if yearning for the light to come back. A third copied a teapot's whistle, and when Engle got a new, non-whistling kettle, the bird would still whistle whenever the pot was placed on the stove. These birds knew more than the realm of sounds surrounding them—they could place the sounds in a context.

These stories suggest that birds' ability to imitate a wide range of sounds may have something to do with a sense of themselves in an environment. This is an idea yet to receive serious study, especially in the wild. Most studies of song learning in captivity have involved isolating birds in different ways and subjecting them to different kinds of controlled stimuli. Valuable, specific information can be gained as a result. But we may miss the larger function of extensive multifaceted song. Social species like starlings seem to use songs in subtle interactive ways that are far more nuanced than attraction or defense.

Donald Kroodsma, recently retired from the University of Massachusetts at Amherst, is one of the world experts on the complexity of bird song, what the field calls the *song repertoire*. Why do some birds—such as catbirds and mockingbirds—sing so many different types of songs? How do they learn this catalog of tunes? Through imitating other birds, either their elders or other species in their midst? How much of the music do they invent themselves? Research has shown that complex bird songs are usually learned by individual birds, unlike calls, which are more likely to be innate. It is odd that birds are born knowing those short sounds with specific meaning, while the more expressive, less functional sounds take effort and practice. No one said music is easy, for birds or for us. Kroodsma admits that after forty years of serious work on bird songs, we know very little about why some of them are so enigmatic and complex. "We still don't know why mockingbirds mock."

Perhaps this is what led Kroodsma to try a famous learning experiment on the catbird instead. The gray catbird, *Dumetella carolinensis,* is a common medium-size North American bird known for its flashy, aggressive presence. He sits on an exposed perch (where the phrase *catbird seat* comes from), imitating whoever he likes, adding snatches of pretty melody, then interjecting short harsh notes in between the chortling phrases. Sometimes these squawks sound a lot like a meowing cat, hence our name for what the Chippewa called *Ma-ma-dive-bi-ne-shi*—the bird that cries with grief. Samuel Harper, author of the intriguing old volume *Twelve Months with the Birds and Poets,* a mixture of diaristic nature writing and an anthology of quaint bird poems, includes the following verses on the curious dual quality of the catbird's song:

You, who would with wanton art
Counterfeit another's part
And with noisy utterance claim
Right to an ignoble name,—
Inharmonious!—why must you,
To a better self untrue,
Gifted with the charm of song,
Do the generous gift such wrong?

Oh! you much mistake your duty,
Mating discord thus with beauty,—
'Mid these heavenly sunset gleams,
Vexing the smooth air with screams,—
Burdening the dainty breeze
With insane discordancies.

I have heard you tell a tale
Tender as the nightingale,
Sweeter than the early thrush
Pipes at day dawn from the bush,
Wake once more the liquid strain
That you poured like music-rain,
When, last night, in the sweet weather,
You and I were out together.

Unto whom two notes are given,
One of earth, and one of heaven,
Were it not a shameful tale
That the earth note should prevail?

Of all the poems in his book, this is the only one he claims is "anonymous," so I suspect he wrote it himself, and didn't quite want to sit out on the catbird seat with it. How else but in verse to express one's displeasure at a music that doesn't come out from the bird quite the way you wish it did? The catbird has its own distinct aesthetic, different from ours and different from the starling's. Your song or mine? Cast aside your prejudices to learn some new rules and structures. So a gifted musical bird wastes his time imitating other species and inserting meows and squawks in the midst of resounding beauty. Maybe he's an avant-garde jazzbird destined to shock the world into something more than pretty melodies. Contemporary poet Richard Wilbur heard the catbird's quirky mix as somewhere in between truth and fiction: ". . . it is tributary / To the great lies told with the eyes half-shut / That have the truth in view"

We aim for the truth of what we hear but when we listen to sounds not meant for us, we eavesdrop with our ears half-closed. What of the catbird behind bars? Kroodsma raised young catbirds in the laboratory, in a situation where their song learning could be controlled. Two groups heard only a repeated, ten-second clip of normal catbird song. Two other groups heard a much longer, sixteen-minute clip of song, repeated over and over. The fifth group heard no taped song at all. Kroodsma expected to discover some correlation between the amount of practice material each bird heard and what it came to sing, which is what had been found in similar studies on other species. Yet something rather surprising occurred. Each bird developed its own unique song largely as a result of what bird song scientists call *improvisation*—making up a song on its own, with no teacher to help. (Since it does involve practice working up to the performance, as a musician I would rather call this *composition*.)

All the birds in the study developed distinct repertoires of hundreds of separate song syllables. Even the birds that heard no catbird song during the crucial learning period were able to create their own distinct songs that later got a favorable response when the birds were released into the wild. The distinctiveness of the song seemed to ensure success, not any particular sounds being copied or assimilated.

What exactly were the catbirds singing out there on their seats? Kroodsma says nothing about the particular qualities of the hundreds of sounds his team identified, only that they were diverse, catbird-esque, and appreciated, regardless of how much training each bird had. They were all creative individualists, driven to their own tunes. Compare this to a study conducted by King and West on brown-headed cowbirds, the bird whose song has the largest known range in pitch, more than four octaves. Male cowbirds raised among canaries only imitate canaries; but surround them with female cowbirds and they will begin to wildly improvise, even though the females make no sound at all. The male changes its song as a response to social, not sonic, stimulation.

So birds extend their repertoires through imitation and invention. But why build a large repertoire at all? Perhaps the longer song is simply less boring. Charles Hartshorne is most remembered as the great twentieth-century theologian and philosopher who argued that God is a process, not a cause, an idea he built on the shoulders of his teacher, Alfred North Whitehead. He spent nearly a century on this project, living until the age of 103. He also was renowned as an ornithologist and wrote one of the finest books on the more abstract problems of structure in bird song, *Born to Sing.* Although Immanuel Kant remarked that we do not get bored or impatient with a bird who repeats the same simple song incessantly, Hartshorne begged to differ. He believed that there is a "monotony threshold" in birds' own aesthetic sense.

Birds with simple songs don't sing all the time. They take long breaks between songs. Birds that sing prodigiously and constantly are more likely to have a complex, ever-varying song. Hartshorne compares two of our common woodland species, the ovenbird and the brown thrasher. The ovenbird chiefly sings a characteristic phrase, *teacher teacher teacher,* only altering the number of times *teacher* is called out. If he were to sing this ceaselessly, listening ovenbirds, as well as humans, might soon get fed up with it. Luckily, he tends to pause at least twenty seconds between songs. Contrast this to the brown thrasher, the one species of bird in the world believed to have the greatest number of specific songs in its repertoire, nearly two thousand: "He hastens on from phrase to phrase, immediately repeating most of them once or twice only. . . . Pauses are so short that one can seldom time them definitely. A bird performing in this fashion does almost nothing but sing." The ovenbird, instead, is "largely intent on other things."

Scientists have tended to note Hartshorne's hypothesis but have not always taken it seriously—it harkens back to Darwin's presumption of an avian aesthetic and would be too hard to quantify. But Kroodsma believes that Hartshorne is right in one crucial sense: It is not the total number of song types that creates complexity, but the *contrast* between one phrase and the next. Variability must be perceived as it goes along, and its constant presence must indicate some evolutionary advantage.

Hartshorne, as a believer in a God who manifests Himself in our world through the eternal unfolding of life, heard something else in the songs of birds that he was not afraid to bring up: bliss.

> Must not singing be enjoyable in itself? Other forms of skill are enjoyed, indeed all forms, I maintain, which are engaged in as freely and persistently as this one. . . . There are many levels of musical feeling, as there are many levels of life, between insects, amphibians, or birds and man.
>
> We know that songs are sometimes sexual stimuli, not only between mated birds but between rival males. This, too, is not absolutely unmatched in human life. The difficulty is not to find analogies but to know how to make them useful, rather than merely misleading.

Though most bird song scientists love these songs as much as less disciplined listeners, they shy away from even asking the kind of questions that intrigued Hartshorne: What exactly is musical about bird songs? Which bird is the best singer of all?

Hartshorne found nearly every conceivable attribute of human music somewhere in the songs of birds. *Accelerando* in the field sparrow and ruffed grouse; *ritardando* in the yellow-billed cuckoo. *Crescendo* in Heuglin's robin chat of Africa; *diminuendo* in the Misto yellow-finch of South America. Harmonic relations in the crested bellbird of Australia and the warblers of Fiji. Themes and variations in the Bachman's sparrow. The closer one listens to bird songs of all types, the more one hears structure, order, pattern, and design. Hartshorne doubts that human composers could do better in the short spans of time birds usually allot to their tunes.

The longer songs do the best job of staying above the monotony threshold. Hartshorne was not the first to rate the relative qualities of different singing birds, but no one else has been as simultaneously aesthetically daring and rigorous. He came up with six parameters of song development, and an equation to weave them all togther: loudness, complexity, continuity, tone, closure, and responsiveness to environing sounds. With precise calculations of these admittedly rather vague characteristics, he created a master list of the greats. How do some of our favorite singers stack up? Out of a list of nearly 200 "superior singers," some of our friends are on the charts:

Superb lyrebird, number 1.
Albert's lyrebird, number 2.
Spotted-breasted laughing thrush, 22.
Mockingbird, 53.
Brown thrasher, 61.
Nightingale, 70.
Thrush nightingale, 71.
Hermit thrush, 114.
Marsh warbler, 129. (Who knew?)

If you're familiar with these birds' songs, you can see that Hartshorne goes for complexity, pattern, and sheer volume rather than delicate poetic beauty. He is not carried away by the lifting lines of the hermit thrush. Like most, he couldn't parse the crazy fragments of the marsh warbler. He doesn't even list the scrawnchy sounds of the starling or the catbird. You might as well be comparing Mahler to Mozart. Whatever your numbers, some part of music remains in the ear of the beholder.

Even if the beholder is a bird—catbird, thrasher, mockingbird—all closely related, with similarly virtuosic songs. Can each bird tell its song reliably from the others? That question turned out to make one of the most important playback experiments ever conducted in the wild.

Bird field guides teach humans to tell these three songs apart using some simple guidelines. The catbird tends to interject surprising, harsh notes into an otherwise melodic explosion, without the immediate repetition of each element. Mockingbirds repeat their imitations in groups of three to seven, often with higher levels of organization: ten- to twenty-second phrases with beginnings, middles, and ends. Brown thrashers fall smack in the middle, tending to repeat each motif twice. They use the greatest number of individual motifs—more than any other bird species yet studied, singing so many different patterns that it has been impossible to tell thrashers by their signature sounds alone. What matters instead is *how* the syllables are presented: catbirds once, mockingbirds three to seven times, thrashers twice. Do the birds themselves use the same rules to tell their similar songs apart?

To find out, Michael Boughey and Nicholas Thompson at Clark University spent several years on an elaborate field project, playing different versions of thrasher songs back to the birds themselves and observing how they responded. They used the original thrasher songs. Then they "catted" them, taking out the repeats. Then they "mocked" them, adding additional repetitions using painstaking re-recording and splicing of tapes—this in the days before computers made sound editing as easy as text editing. They also "thrashed" catbird and mockingbird songs, creating an artificial sense of doubling. They took the tapes and an amplifier out into the field, and once they heard a singing thrasher they would play the various versions of the songs to see which one it liked best. "*Liked* is perhaps the wrong word to use," they reported; "*disliked* might be better." A higher-rated response was one when the bird came close to the speaker and sideswiped it the way fighting thrashers brush each other during territorial disputes.

The scientists were shocked that it actually worked. Thrashers came closer and more often to the speakers blaring songs with the characteristic twoness. But the means behind that double quality was not so simple—the birds don't *always* repeat each syllable twice. Sometimes they repeat them three times, sometimes not at all. Previous listeners, more mnemonic than scientific, described the song thus: Thoreau at Walden Pond—"Drop it, drop it—cover it up, cover it up—pull it up, pull it up, pull it up." Our friend Schuyler Mathews: "Shuck it, shuck it; sow it, sow it; Plough it, plough it; hoe it, hoe it." And a Mrs. H. P. Cook, in 1929, imagined the thrasher at one end of a telephone conversation: "Hello, hello, yes, yes, yes, Who is this? Who is this? Well, well, well, I should say, I should say, How's that? How's that? I don't know, I don't know, What did you say? What did you say? Certainly, Certainly, Well, well, well, Not that I know of, Not that I know of, Tomorrow? Tomorrow? I guess so, I guess so, All right, All right, Goodbye, Goodbye."

Boughey and Thompson were puzzled that the sense of twoness, which the thrashers so clearly noticed and responded to, was not exact. Why shouldn't a simple rule govern their behavior? Wouldn't that be much easier? All right, if they didn't repeat each syllable twice, did they sometimes break a longer phrase into two equal parts? Sometimes. *ShuckitShuckitSowit—PloughitPloughitHoeit.* Did they sometimes follow one phrase with a different one, but a difference similar enough to be considered "two" of *something*? No. Perhaps the whole search for a single parameter to define the thrasher's song was barking up the wrong tree.

Yet if a birder hears a thrasher, after ten or twenty seconds she can tell it is a thrasher. The twoness can be heard. The average number of repetitions is two, even though it isn't two every time. Boughey and Thompson concluded that the birds must listen to a longer strain of song before knowing that it is one of their own. Perhaps this is why their song is so long, embellished and variable, because they only recognize one another upon such a law of averages. Forty phrases per minute, twenty minutes at a time.

A bird that sings continuously without pause needs to keep varying its song or would-be listeners will lose interest. It appears to confirm Hartshorne's monotony hypothesis. *Maybe, maybe. Okay, okay, okay.* There is still no clear reason why any species would need to work so hard to hear the song that is meant only for it. Nature does not always choose the simple road. Complexity may have some value we are not yet able to see.

This is no surprise to anyone who has tried to understand music. Are there really "rules" that govern the smooth motion of harmony, as they teach you in college classes? Once you learn these rules, you gain a new appreciation for the music of Bach, who was able to write wonderful chorales while staying wholly within them. As any composition student knows, it is nearly impossible to write something that sounds good while following all the rules; if you're not Bach, you have to break them quite often to get the point across.

If birds are making music, this is what we would expect to find: There are rules that *suggest* the form and content of the songs, but these rules do not enable us to automatically predict the songs. Music presents an idea that only stands for itself, that cannot be released into the air any other way. The songs themselves cannot be reduced to formula, but they may arise as an order that emerges from simple reiteration, like the fabulous patterns inscibed into seashells.

Perhaps mimicry is the easiest route to complexity: If a longer, more extensive song is more attractive to females, then a quick way to get such a song is to copy what you hear around you—beats having to make up a completely original songbook! But if all you do is sing everyone else's songs, how would your species stand out? How can listeners tell you're the starling imitating the cardinal and not the cardinal himself? It would have to be the *way* the imitations are combined. Mimicry is never just plagiarism; instead, it's using all the sounds that you hear in very species-specific ways. That's why marsh warblers always know they're jamming with other marsh warblers, enjoying their unique ability to sing cross-cultural Afro-European bird world music. Mimicry in full, adult bird song is part of a species-specific songmaking strategy. Or style—depending on whether you hear it as art or as a game.

I am happy to discover that no rule is the ultimate rule. I am ready to listen and be amazed by how much always remains in the bird world to sing along with and to decipher. Learning is a mixture of imitation and invention. The birds too are tossing and tumbling among one another's sounds. Some birds are taught the way to do it by their elders, but others just listen to their whole environment as an exciting, pulsing soundscape to tease with.

The marsh warbler's song is so far away from monotony that it makes one dizzy to even contemplate dissecting it. How do these little birds learn how to sing so much and so well, copying all the surrounding sounds that they can? No one knows much about it save Françoise Dowsett-Lemaire. This is what she told me—"As for fathers teaching their sons: no, this does not happen, because they stop singing once they feed. Many males even stop singing several days before they start feeding, and some stop as soon as they mate." So the little birds, out of the nest at eleven days old, must immediately start listening to the sounds around them and start shaping what they hear into their own relentless ultimate song. None of their own kind are around to teach them, as their elders remain mute.

Must the baby warblers form their song on their own? I always wondered how they fabricate the complicated motifs in which they alternate imitations (1-2-3-4, 1-2-3-4, etc.): Is this copied or invented with each generation? The answer to all this would have been to raise young marsh warblers in soundproof chambers and test them with various motifs week after week, from other species as well as from their own. Ah, another fine playback experiment to be tried. But then we would have the bird in controlled circumstances, no longer the resounding habitat of its own wild world. The magic of the marsh warbler is that it is a thorough appraiser of its own environment. From Northern Europe to East Africa, alighting by the Mediterranean along its way, he picks up the songline of migration, an ultimate triumph of imitation *and* invention. We're not likely to succeed at slowing this bird down, but we know a little bit about his style. Starlings, catbirds, and thrashers have equally distinctive musics, but few people have taken time enough to hear them. Deep listening to the shape and the form of a bird song this complex is as difficult as it is rare.

CHAPTER 6

Rhythm and Detail

RESEMBLANCE BRINGS the first glimmer of understanding. I know this song, it sounds like something I've heard before. No wonder mimicry was the first aspect of bird song structure to receive close attention. Delving deeper into song's complexity requires years of close listening and observation, in the fickle wild, not the bounded cage. Those who have risen to such a challenge have been obsessed with their subject in the best sense. Through meticulous work they struggled to learn how it feels to sing like a bird and know the rhythm and detail of the song itself.

No student of bird behavior has been as diligent as Margaret Morse Nice of Columbus, Ohio, who spent eight years in the 1930s precisely recording the movements and activities of all the song sparrows on the flood plain behind her suburban house, a forty-acre tract that she christened "Interpont." With a minimum of equipment and a maximum of enthusiasm, she pursued a quandary that possessed her ever since she studied with Konrad Lorenz in Vienna: Were birds like machines who responded instinctively to their surroundings, or were they capable of planning their actions? She aimed for what philosophers call a *phenomenology,* paying direct attention to the happenings you see with as few preconceptions as you can make do with.

Nice learned every inch of her backyard study area and how it was divided up by her subjects. She mapped out the territories of each bird, putting bands on their legs so she could tell them apart. Each was given a number-and-letter code name, with all their comings and goings dutifully described. All kinds of variation in behavior were observed. Some of the birds migrated and left each winter, others did not. Most chose different mates each season. One male even mated with his sister for a time. Her detailed report of all this activity fills two volumes, more than 500 dense pages. No comparable study on a single bird population has ever been published before or since.

Song is vital to song sparrows—their Latin name is *Melospiza melodia,* meaning "melodious singers." Mathews, in his field book of transcribed bird songs, considered them the "best exponents of the song motive among all the members of the feathered tribe." Saunders of the squiggly neumes praised the pleasing and attractive nature of the songs, which he found more persistent and frequent than those of any other common songbird. Song sparrows begin with one declamatory rhythmic outburst and then move on through related rhythms in a microcosm of two seconds or less. The phrases are intelligible and clear, often resembling familiar themes from human music. Nice heard the beginning of Beethoven's Fifth. Mathews heard Verdi's *Rigoletto:* "*La donna è mobile!*" Or "wail, wail, fickle wife is she, flown away and left me!" "Sad, sad, what a tale of sorrow! She may return tomorrow." Or in bird words, *Wertz, wertz, wertz, weet-weet-weet-weet-weet-weet spee-ge-wee-ge-dee.*

As in the song of the thrasher, each phrase is a little bit different, but they all have an identifiable song sparrow essence. The males Nice studied had between six and twelve songs. Saunders classified more than 800 records of song sparrow songs and found that most males have either nine, twelve, eighteen, or twenty-four song motifs, strangely enough often in groups of three. For the three males that nested right in her garden—1M, 4M, and 187M—Nice wrote down all their phrases. 1M had six songs, including *chip chip chee yer zig zig*

zig zig; chee chee chiddle hair terpée terpée terpée; and *tee tee tee eeeyer huffum huffum huffum.* 4M had nine, among them *spink spink spink spink ereteree* and *hur hreeeee tweet tweet tweet tweet tweet.*

Nice knew of Saunders' notation but found it too technical and idiosyncratic for her purposes. Song sparrow songs are supposed to have a three-part structure, beginning with a few rhythmic notes, then a series of trills, and more unpredictable ending notes. She heard much more variation than that. For 288 hours she recorded all the singing done by those three birds closest to her home. Her book reports all kinds of specific conclusions in paragraphs like this one:

> A song sparrow usually goes through his whole repertoire before repeating any one song; the order in the second series is rarely repeated exactly. A bird with 6 songs gives 2/3 of them in an hour in the inhibited state, all in the uninhibited state, all and half again in the stimulated state, and twice over in the highly stimulated state. A bird with 9 songs will present them all in the stimulated state and do the same with a start on a second rendition when highly stimulated. I do not know what pattern is followed by a bird with 24 songs.

There are hundreds of pages of data such as this, with few of the complex statistics favored by more recent bird song science. Nice is more interested in the quality of observation than the quantity. She would rather elucidate than enumerate.

Despite her desire for uninhibited observation, Nice was obviously impressed by the latest ethological theory of her day—namely, Niko Tinbergen's assertion that bird song is much more about defending territories than about attracting mates. This fits her observations: male song sparrows sing most intensely when showing off to other males, announcing their boundaries, defending their zones. Singing cools down but still continues after the eggs are laid and the females are guarding them on their nests.

> While his mate is incubating he gives a "signal song" for her to come off the nest—a sign that all is well and he is ready to guard. He sings

> near the nest while she is off, a warning that he is ready to drive off intruders. His singing while she incubates may be an expression of satisfaction that all is going well, a method of passing the time when he is alone. While caring for young, there is little energy for singing, except for single songs after feeding the family, and especially after carrying off excreta. The fall singing in fine weather would seem to be an expression of excess energy.

Singing in the sun just to let off some steam? Nice quotes the great biologist Julian Huxley's view that bird song goes on beyond its functions. "Song," says Huxley, "is simply an outlet, and a pleasurable one." Birds "continue to sing in all moments of excitement or exaltation." Exaltation? Is that something a scientist can determine? Nice's birds turn out to be more than machines. They are animals with excellent memories—of territory, of songs heard, homes to return to, and strong emotions—*feelings* explained by a mixture of instinct and learning.

Consider the various "methods of intimidation"—in both sound and stance—that Nice observed in her song sparrows. Tall, erect posture, for the full advertising song, announcing "here I am, this is my place. These are all the songs I know." Antagonism, with an open bill. Menace, with lowered crest, crouched body, extended neck, beak pointed at the enemy. Threat posture, with feathers ballooned. The challenge—"puff-sing-wave"—wings vibrating, slow hovering flight in place. Finally the intriguing courtship behavior of *pouncing,* when the male darts down, nicks a female, and flies off with a song, aggressively, letting her know he's interested. Mating itself doesn't come until weeks later.

Nice's favorite singer was 4M, whom she observed very closely through his life of many years beginning in 1928, charting his series of different mates and skirmishes with neighboring rival males. She recounts his dramatic final season at Interpont: By 1935, after years of successful broods, 4M was having trouble finding a mate, and finally ended up with one that Nice didn't like very much, so she left her scientific pose to name her after Socrates' shrewish wife. This female bird was a "cold old-maidish creature, tyrannizing over her fine husband like a veritable Xantippe." She often made secret forays over to a nearby male, 225M, and 4M often had to run over and nudge her back home. She was lazy about nest building, and when she finally did, she laid only two eggs. Some wrens came by and poked them, and eventually Xantippe left. Mrs. Nice said good riddance.

Five days later, on May 11, 1935, 4M did something that Nice had never observed in all her song sparrow studies. Beginning before the sun came up, he put forth an incredible outpouring of song that kept on all day until after the sun went down. Beginning at 4:44 A.M. with song D, he launched into his cycle, with five songs a minute, two hundred per hour, on and on tirelessly through the day. Grief? Longing for a new mate? Relief? Remorse? "Excess energy"? What was it all about? He did not find another mate but kept singing in a more guarded way for a few more weeks. Late in summer 4M departed. He never returned to Interpont again. Nice had watched him find eleven mates, build seventeen nests, and raise thirteen chicks. He was widowed seven times.

How do Nice's conclusions compare to more recent studies of the song sparrow? After all, she did not have tape recorders or sonograms to back up what she heard in the songs, only her ears, eyes, and time on her side. She conducted no playback experiments to test birds' responses to different variations of their own songs. She relied entirely on her powers of discrimination—might the birds have their own criteria of the same and the different? Perhaps what we consider to be distinctions in song type really don't matter to them. Maybe they hear other variations we cannot.

Song sparrows are excellent birds with which to investigate such questions. They thrive among humans and are somewhat tame and easy to watch. They have a complex song, but nothing that taxes our listening abilities like the starling or the marsh warbler. Each song sparrow has at most thirty song types, and they repeat them in a regular order. This is much easier to track than the extensive performances of catbirds and nightingales. Song sparrows also share some of their repertoire with their neighbors, and this sharing can be quantified and measured. Among males with larger repertoires, there is greater similarity between the different songs. Unlike thrashers and mockingbirds, song sparrows cannot learn indefinitely.

In the eary nineties a group of scientists, led by Peter Marler and Jeffrey Podos, tried to divide sparrow song motifs into the smallest unit of organization, either single notes or groups of notes that always occur together and in sequence. They called these "minimal units of production," which sounds more like an industrial term than a building block of music. Through playback experiments, they found that males react more intensely to differences *between* one song type and the next, rather than to subtle variations of the same type, something that Kroodsma had previously suspected. The scientists were sensitive to the units of production, but the sparrows weren't. Rather than categorize their own songs as being composed out of building blocks, they were more interested in change from one song to another.

Another experiment showed that if you play a song sparrow one of his neighbor's songs, he will not reply with the same motif but with a different riff from the list that he shares with that neighbor. The researchers called this *repertoire matching,* as opposed to *type matching.* (Scientists often invent or redefine their terminology with each new article in the hope that their colleagues will follow their new terms. Often they don't.) But if the male sparrow hears a completely alien song from a stranger sparrow, not a neighbor, then he will try to match that song with the closest type that he can! Whatever for? Some have concluded that matching with a similar type of song is somehow more aggressive than trading common phrases with the neighbor bird. It's like two jazz musicians meeting on the stand. One is soloing over "My Favorite Things," and the other starts to jam over the same chord changes at the same time. That might suggest a contest. But if the second player says, "Oh, I know that guy. He loves the tune 'Summertime,'" and then switches to "Summertime," it's a friendly response, demonstrating respect, not a need to win. Song sparrows seem to recognize these two ways of singing together.

The team tried a third experiment. If copying is more aggressive than riffing, then it should be more prevalent early in the season, before mating, when birds are spending more of their time in aggressive territorial defense. Seventy-three percent of the time this turned out to be true. Whereas Nice and Saunders heard magic groupings of songs into 6, 9, 12, 18, and 24, researchers armed with the latest technologies did not find playlists divisible by three. But they did get closer to the birds in one way that earlier observers did not, by showing that what mattered to the singers was something that did not so much impress human observers looking for basic rules. Where people looked for similarity, the sparrows wanted difference. Podos and Marler found this out by playing their own songs back to the birds and watching what they did.

The use of playback as a technique in bird song study is clearly one of the great advantages sound recording technology has given the scientist. Yet science itself has expressed some serious reservations about the technique. Don Kroodsma was the first to point out this danger: If you want to see how a bird will respond to an unfamiliar song or sound, you can play him or her the sound. The first response will be to something unfamiliar, but pretty soon your subject gets habituated to the sound. Judging the relevance of the reply gets complicated. What did Kroodsma suggest? Don't just use a single playback sound, but a series of playback sounds that are slightly variable. Then you might be able to avoid what he calls *pseudoreplication*—testing for one thing while actually getting a result for something else. The bird may get too used to your tapes and no longer treat them as something worth responding to.

The song sparrow experiments reviewed above were all done after Kroodsma's critique, and in a later review he praises them for the care with which they were conducted. Nevertheless, they are still artificial situations. They test for specific hypotheses in constrained conditions. They are not the results of disciplined observation of how birds actually live and sing in the real world. Their conclusions will always be statistical and tentative.

Playback experiments may be closer to art than to science, nearer to interspecies music than rigorous test. Play a clarinet to a bird and listen to what happens. From the first note I'm messing with his sonic world. I don't want to prove anything, I am only trying to forge a musical link. Trying to learn from the bird's ways without so simple a model as me copying him or him copying me. There are many more possible reactions than that. The song sparrow story gives me hope: I want to share exuberance, adventure, and some common cause of music for its own sake. Perhaps the bird may learn that he and I like the same kind of songs.

But remember that each species is unique, with a particular musical culture. We cannot quickly generalize from this kind of sparrow to any other bird. Consider instead a long solo song, sung on and on in clear patterns. The longest monograph on a single bird song is not about a complicated learned song, but instead a series of brief melodies composed out of three notes, probably innate and not learned at all. In 1943 Wallace Craig published a special issue of the New York State Museum Bulletin entitled "The Song of the Wood Pewee . . . : A Study of Bird Music." It is nearly two hundred pages in length.

The eastern wood pewee, a small gray forest flycatcher, sings only three simple phrases, combined in a series of clear groupings, easy to note and identify. Why did Craig choose to devote so much attention to so simple a song? Because it sounded pure and beautiful to his ear. None of the endless variations of the thrasher or mockingbird; no complex sound-matching society life like the song sparrow. What was so musical in this delicate pewee clarity?

Craig gathered twenty-two observers from across the eastern and midwestern United States. Together they amassed 144 records of the morning twilight song, a total of some 93,000 phrases. What an incredible amount of pewee music to consider! What did they find? The wood pewee's morning song lasts between 16 and 32 minutes, with an average of 24. The typical song contains 750 phrases, the longest, 1,273. The song follows a definite pattern among the three possible phrases. At any one moment there is a tendency to move between one pattern and another, like those thrushes whose motifs follow each other according to a Markov chain. Craig sees not a mathematical tendency but a musical one. Like true music, the song sings of nothing beyond itself. "While singing the twilight song, the pewee is more or less isolated from the practical world."

To catch the very beginning of the pewee's twilight song, you must arrive in the solid darkness of late night. First the bird begins with one sleepy call. Then silence for several minutes. The sky imperceptibly lightens. The wood pewee, says Craig, is attuned to the early morning changes in the forest, the trees, and all of its other inhabitants. Gradually the singing becomes more intense, even incessant. The three phrases are repeated one after another, in all manner of combinations. Here is how Craig identified the early song's components:

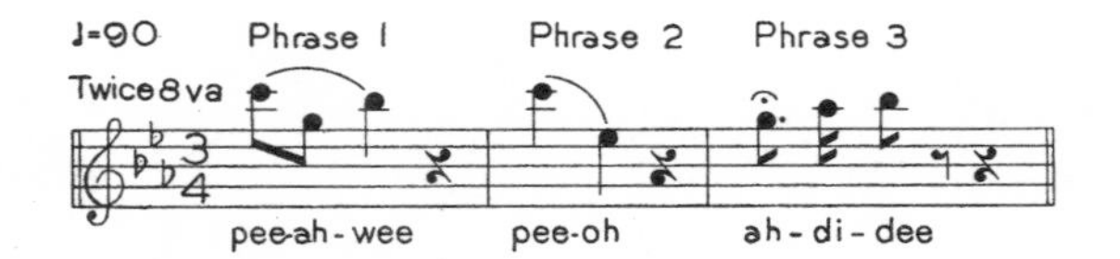

The three phrases of the morning twilight song

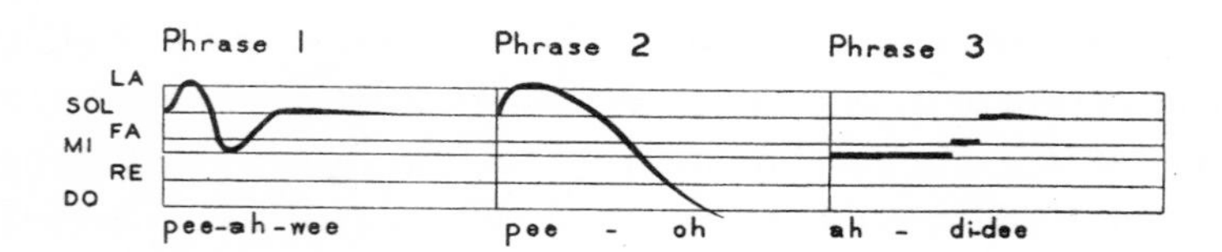

Graphic representation of the three phrases

The sentence 3132, the wood pewee's most perfect sentence.

Record of morning twilight song by Saunders's Thirteenth,
Fairfield, Conn., June 25, 1932.

(Arrived at 3.24 a.m.) (About 3.44 a.m.) 1 1 1 12 312 312 132
31312 312 3132 31312 3132 (31312 twice) 3132 (31312 four times)
15th 36th 50th
3132 (313132 twice) (31312 twice) 3132 (313132 eight times) 31312
74th 86th 100th
3132 313132 31312 3132 (313132 six times) 3 1 3 312 3 1 3 312
157th 172d 208th
3132 (313132 eleven times) 1 . . . 2 3 1 3 3 2 13132 (313132 twice)
224th 293d 302d
3133132 (313132 twice) 3 1 3 312 3132 (313132 twice) 3133132
314th 321st 343d 355th
3132 313132 . . . 1312 (3132 twice) 3133132 3132 3133132 3 1 3
384th 395th
312 3132 3133132 3132 3133132 313132 3 1 3 3 2 1 3 (3132 twice)
412th 423d 437th 443d
3133132 (313132 twice) (3133132 twice) 313132 - 1 3 1 - 12 3132
451st 458th 470th
3133132 313132 3133132 3132 (3133132 thrice) - 1 3 312 3132
499th 512th 523d
(3133132 twice) 3132 (3133132 twice) 313132 3 - 1 - 2 313132

PARTIAL RECORD OF PHRASES IN
ONE WOOD PEWEE'S DAYBREAK SONG

Craig even employed Aretas Saunders, whom he considered the nation's greatest expert on bird song, to write down pewee songs for him. Above is the first half of what Saunders heard on June 25, 1932, starting at 3:42 A.M. You see anything but randomness here. The most common pattern is what Craig calls "the wood pewee's most perfect sentence"—3132 as composed of the numbered phrases above. "Are they gone? I don't know. Are they gone? No." That's mnemonic language, but Craig hears music. He wonders what it would take to understand the musicality of this balanced, complete sentence. Here's what he decides: Phrase A is a fragment, a musical question, looking onward for an answer. Section B_1 pushes on toward an end, but at the last moment turns up. The question again, and with B_2 a more complete, descending, landing answer. AB_1AB_2—a popular human song form, first noted in the pewee by Henry Oldys in 1904, who compared it to "Way Down Upon the Swanee River." Tell that to the starlings, eh? Craig believed that the "perfect" 3132 sentence is the most highly evolved form of bird music known to human ears.

Craig concluded that the rhythmic song of the wood pewee, sung long and leisurely through the earliest morning hours, is a true music, not just an outpouring of raw emotion. The pewee's whole demeanor while singing before dawn expresses calmness and lack of excitement. The bird sings in the dark and is wholly wrapped up in the song. He is as obsessed with his own music as those twenty-two human listeners who listened for many shadowy hours to take it all down.

There is no other scientific study with the stated aim that "our chief interest is in bird songs as *music.*" Craig believed most previous bird song studies were too concerned with the function of song and too little with the qualities of the song itself. The situation hasn't changed much since—only Hartshorne's *Born to Sing* and the book you are now reading are the heirs to Craig's unique direction. Who was Wallace Craig and how did he end up seeking music in such an unlikely place?

Craig's most famous work is a 1918 paper called "Appetites and Aversions as Constituents of Instincts." This is one of the most coherent early arguments against the idea that animals are simply reflex machines, responding automatically and predictably to the same stimulus every time. Instead, Craig postulates a view of birds as living through a series of overlapping cycles of behavior. Eating, defending, mating, singing—each cycle competes for attention and has its own appetities to be satisfied and fulfilled. The bird's attention waxes and wanes from one to the next.

By the time he took on the wood pewee, Craig had been studying bird behavior for more than thirty years. Through his work on doves and pigeons, he was an early proponent of the view that birds, like other animals, were far more complex than earlier ethologists had thought. Margaret Morse Nice had introduced Craig to her teacher, Konrad Lorenz, who wrote in his autobiography that "Wallace Craig became *my* most influential teacher. He criticized my firmly held opinion that instinctive activities were based on chain reflexes." Craig convinced Lorenz that organisms do not react automatically to the same stimulus each time it is given. Instead, as soon as the stimulus appears, the animals begin at once to actively seek the situation for release. Still mechanistic, but not blindly so.

Lorenz is most popularly remembered for the notion of *imprinting* that leads baby geese to follow their keepers around, believing them to be their mothers. He also presumed that the wing spots on ducks came in such odd colors not because they were beautiful, but because they were the most *improbable* colors, perfectly tailored to generate the appropriate reaction of species identification in the receiver. So, wrote Craig to Lorenz in 1940, is this the same for the specific musical intricacy of bird song? Is there no more to it than improbable accident? Here's how Lorenz responded:

> I am very far from interpreting everything as a releaser and I have begun to have my doubts about the releasing function of the *details* of bird song. . . . It is certainly more beautiful than necessary and in this is akin to human art in general. Art *is* a fact and after all it would be rather ridiculous from our evolutionistic ideology to deny the possibility that something similar may occur in other species

There you have it, from the master ethologist himself: art is a fact. Craig still firmly believed instinct guided the pewee's song, and it is probably true that the young pewee, as one of the flycatchers, may inherit his song and not need to learn it from adult tutors. Craig's methodology was listening, and he worked hard at it, even claiming to hear the process of evolution at work in the pewee's song. Craig believed that the specific three-part song structure evolved because it is "musically suitable for singing in a continuous rhythmic song." He believed that the *tendencies* toward these phrases were older than the phrases themselves.

Phrase 3, says Craig, evolved to continue for many minutes and hundreds of repetitions. Musically it seems to reach onward and upward. Phrase 2 has finality but no clear rhythm. It evolved for leisurely daytime singing, which came first in the bird's development, later incorporated into the more intense predawn song that is unique to this species. It is also sung by the *western* wood pewee, which Craig considers an older species because it does not have the distinctive and more stylized morning song, lacking phrase 1. Phrase 1 is a compromise between the other two phrases, the result of a penchant for quick tempo and something akin to Hartshorne's monotony threshold—revealing a preference for delicate balance between repetition and novelty.

A pewee can *ah di dee, pee a wee, ah di dee, pee oh* all through twilight, leaving no listeners bored. The song is designed to resound on and on. This bird has evolved a propensity to continue, to sing the sun up in the earliest hours of dawn like Orpheus with his lyre. In this generating motif of an ancient music, Craig hears proof of an aesthetic principle as part of natural selection. He's the only one who has investigated what Darwin imagined was there.

The main line of bird song science, if it acknowledges Craig at all, sees him as a curiosity. "This whole field," said Don Kroodsma to me, "is full of people pushing their own pet theories." They cite the evidence that supports their views and ignore ideas that are too tough to prove. Craig heard music where other scientists heard *units of production* and *packages*. Aesthetics, he reminds us, is far more than prettiness. It is unassailably the result of evolution. Bird song is music not for us but for birds, and don't expect it to be encapsulated by biology. Future bird science ought to consider musicality in its investigations of song. Kroodsma laments that so few scientists have agreed. He said, "I've studied bird song for more than forty years, but I don't know a thing at all about music. Perhaps it's time to change that."

On the other side of the Atlantic, there was one man who tried. In the middle of the green forests of Finland, the zoologist and insect specialist Olavi Sotavalta came across a copy of Craig's voluminous pewee report and recognized its challenge. Sotavalta wondered if he might try out the approach of musical analysis on a genuinely complex bird song, that of the thrush nightingale, *Luscinia luscinia,* an

THE NIGHTINGALE

eastern European and Asian bird with a more rhythmic and scratchy song than the *Luscinia megarhynchos* nightingale of western Europe's romantic poetry. English nightingales sing fifty to two hundred different phrases with much variation and change. Sotavalta noted that thrush nightingales sing an equally large number of distinct phrase types, but each type has a fairly consistent structure more stylized than the phrases of the more famous bird.

Sotavalta was a rare breed himself—a zoologist gifted with perfect pitch, thus uniquely qualified to transcribe what he heard with no special technology. In the 1940s he compiled a list of the different wingbeat frequencies of all the flying insects he met, just by listening to the tones zooming by. He trained himself to distinguish the fundamental pitch from the overtones. His list of these frequencies is praised for its accuracy and still used today. Cornell entomologist Tom Eisner remembers hearing Sotavalta lecture at Harvard in the 1950s, and said that he resembled an Old World monk: tall, bearded, dressed in a long cape like a character from another age.

Inspired by Craig, Sotavalta listened intently to the nightingale's song. Its timbre is not easily harmonious, but raw and complex, combining percussive rhythms and clear notes: "Pure tones could be whistling, piccolo-like, dull, like a low flute, metallic, celesta-like or chippy, like a xylophone, long or short." He struggled to put it in words. "The commonest noise-type appeared in the cadence and resembled the rattle of a tambourine."

The most salient quality of the thrush nightingale song is a series of specific pulses, a general pattern of which pervades each phrase the bird sings, which Sotavalta calls a *period,* synonymous with Craig's *sentence.* I'll call it a phrase, to keep the sense of the bird's whole performance as one long song. Sotavalta studied two birds—one in 1947 and the second in 1948. The first had fifteen basic phrases, and the second, seventeen. At the level of the phrase, a definite form can be identified. In the thrush nightingale song the rhythm seems more significant than the pitch. Here is the basic structure Sotovalta identified that fit nearly every phrase of both birds:

The introductory notes are one or two soft whistling tones. Then a low-pitched antecedent, then a brief link to the characteristic motif, which is the part most distinct between one phrase and the next. Double or triple time, sometimes distinct wide intervals. Then a postcedent series of repeated low notes, a high *bleep,* final "chippy, xylophonelike chords," and that one quick tambourine-type rattle. *Chhuum.* Thrush nightingale deciphered? At least some structure found. Below are four of the fifteen basic phrases Sotavalta heard from one of the birds. The revelation of a drumbeat music more resembles a battery of percussion than a luminous turn of melody. With all the praise given to the nightingale's virtuosity, it is amazing how weird its music looks and sounds. Compare this wild shaking and shifting to a more modern sonogram of several phrases from the same sort of bird, this from a recent study by German biologist Marc Naguib (see next page).

FOUR OUT OF THE FIFTEEN PHRASE TYPES OF SOTAVALTA'S THRUSH NIGHTINGALE

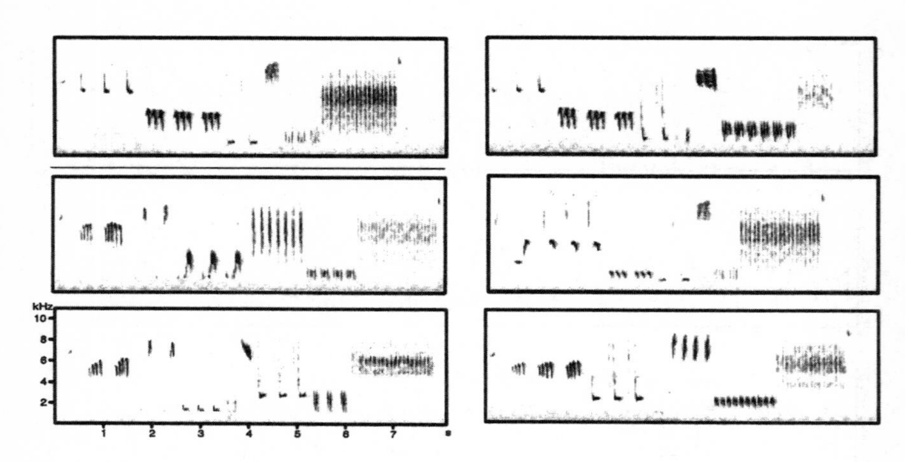

NAGUIB'S THRUSH NIGHTINGALE PHRASE SONOGRAMS

In this case the musical notation shows more nuance, even if it looks too human. Sotavalta analyzed the sequence of phrases in detail and concluded that the bird went through his phrases in a loosely patterned order. There were cycles upon cycles in the structure of the whole song, but each series has some variation. There was a sense of a regular progression through the repertoire, but no sequence of riffs was precisely the same as the next (as the jagged line on page 133 shows).

Sotavalta listened acutely and perceptively to decode the structure of the thrush nightingale's song. He found clear rules in it, yet no line of research was based on his conclusions. Like Nice and Craig, he was a maverick listener, off the main track of sonography, playback, and enumeration. Later nightingale researchers scoffed at his sample size of only two individual birds! And he used a kind of musical argumentation that is difficult to quantify. Yet he traced the secrets of nightingale music more accurately than anyone since. He revealed perhaps why so *little* Western music has actually been based on nightingale song, despite the vast metaphorical power of this bird's image. But the nightingale has been revered outside science, all over the world.

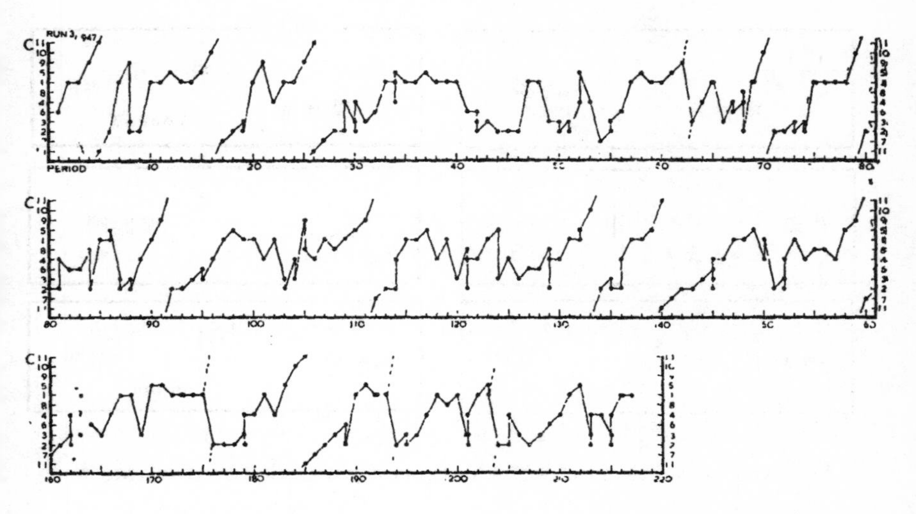

SOTAVALTA'S GRAPH OF SUCCESSIVE SEQUENCES OF PHRASE TYPES SUNG BY THE SAME THRUSH NIGHTINGALE

Nightingales have long had a central role in the musical mythology of Persian culture, within the boundaries of the modern countries Iran and Afghanistan. It is the bird of a thousand stories, *hazâr dastân,* singing turn by turn, *rad bâ rad,* always changing its song. Calling a musician a nightingale is the highest form of praise—the greatest often have the word *bolbol* added to their names as an ultimate honor. In less fundamentalist days, when music was not chastised or banned, bird song was considered a form of *zikr,* or remembrance of God, like a muezzin's prayer. The meaning echoes more in the repetition than in the words themselves. All bird species have their own *zikr,* all praising Creation, and the *bolbol* is the master bird who never repeats himself, always coming up with new names for God. This gives bird song the highest honor in a devotional culture, a loftier purpose than biology has so far allowed.

Despite this reverence, Afghan musicians have not made much specific use of bird song in their melodies or forms. John Baily, one of

Europe's greatest authorities on the musical culture of Afghanistan, brought a recording of English nightingale song and played it to some Afghan refugee musicians living in Pakistan in 1994. They were immediately excited. First they responded to the taped bird song using the "drum language" of spoken *bhols,* in which players speak the patterns they later play on the tabla. Although no one had noticed it before, the birds' phrases fit right into the sixteen-beat recurring *tintal* cycle that is the most popular 4/4 rhythm in that part of the world. *Dha Ti Ta Dha / Ti Ta Dha Ti / Dha Dha Ti Ta / Dha Dha Tu Na.* Then they got out their tabla drums and *rebab* violin to jam along with the tape. To the drummers, the nightingale's phrase was a fully stuctured tabla solo, easy to assimilate and respond to. But their tradition had not explicitly made use of nightingale rhythms before. The end result sounds like a new kind of interspecies music, part nightingale—with the relentless call-and-response not trying to go anywhere or conclude—and the musicians caught in the web of the challenge, trying to play exactly what is heard and to take it to some other, human level.

In Iran, in the Persian music tradition, there is a kind of musical ornamentation called *Tahrir-e Bolboli,* where singers and their accompanists imitate one another with rapid trills and nightingalelike quips. Here is a tale about one of their most famous singers, named Qamar:

> Once upon a day Qamar went to Darband, a scenic place near Tehran, to take a walk and practice in the open air. Qamar started singing Tahrir-e Bolboli while she was walking among the trees. A nightingale sitting on a branch heard her beautiful song and began to sing along. The nightingale was trying to sing like Qamar, and Qamar was trying to sing like the nightingale, just as singers and players meld together in traditional Persian music. The fever rose as they each tried to sing faster and louder. Suddenly the nightingale fell down and died, because it could not keep up with the great Qamar. Qamar cried deeply for two days. She could not forgive herself for having killed a bird with music. Was all this beauty and intensity nothing more than a

fight to the death? Song, whether coming from birds or from humans, must be more than war.

The yearning of the nightingale figures prominently in the famous Sufi fable of Attar, *The Conference of the Birds,* among the most known works in all of Persian literature. The master of birds, the gaudy Hoopoe, is trying to assemble all the other bird species to join him on a quest for the sacred valley. Here's how the nightingale answered the call to join up:

> The amorous Nightingale first came forward almost beside himself with passion. He poured emotion into each of the thousand notes of his song; and in each was to be found a world of secrets. When he sang of these mysteries all the other birds became silent. "The secrets of love are known to me," he said. "All night I repeat my songs of love. Is there no unhappy David to whom I can sing the yearning psalms of love? The flute's sweet wailing is because of me, and the lamenting of the lute. I create a tumult among the roses as well as in the hearts of lovers. Always I teach new mysteries, at each instant I repeat new songs of sadness. . . . If I am parted from my dear Rose I am desolate, I cease my singing and tell my secrets to none. . . ."
>
> The Hoopoe replied, "Although the Rose is fair, her beauty is soon gone. One who seeks self-perfection should not become the slave of a love so passing."

In Persian music and literature, and in Baily's Afghan experiment, we see that much of the musicality of bird song lies in its special use of rhythm as much as its organization of pitches and legible melodies. I doubt it is an accident that we hear these sounds as being closer to music than to words. If birds are "emotional creatures with good memories," as Margaret Morse Nice concluded, then they have what it takes to be good musicians. Sotavalta transcribed a music that people had rhapsodized upon for centuries. He found tendencies toward order, not exact memorized syllables always repeated alike. It was enough for him to find real music in the nightingale's song, without needing to know what it is for.

The serious, controlled experiments that were done on nightingale song in the 1980s and 1990s do not focus on the *content* of what is sung, but instead on what can be more easily tested: stimulus and response, singing and countersinging, how birds react to one another and to playback of their own and their neighbors' songs. Dietmar Todt and Henrike Hultsch have studied nightingales in Germany for decades, both in the wild and in captivity. Because of their work and the work of their students, more is known about the singing behavior of these famous birds than of any other species with so complex a song. Their first studies focused on how the birds sing in the wild, while later experiments examined how the birds learn to sing in controlled circumstances.

One of the first aspects of the nightingale's singing behavior that they uncovered is that there are three distinct ways nightingales sing and countersing to each other, beginning late at night and ending by dawn in the first weeks of spring. Adjacent male nightingales tend to sing back and forth to one another, timing the beginning of each song phrase in a precise way. Most males are "inserters"—meaning that they wait about one second after a neighbor's song finishes before starting their own. Songs alternate between one bird and another. Mutual listening occurs, and timing is everything. Then there are "overlappers," who start their song about one second after their neighbor begins, as if to cover up or jam the neighbor's signal. It's some kind of threat or a mask of the first song, cutting into his airtime. Then there are "autonomous singers," who sing and sing according to their own schedule, paying no heed to what any nearby nightingales are doing. The top bird, soloing without peer? Not a care in the world?

When the scientists adjusted the amount of silence between playbacks of stimulus songs, the inserters adjusted the amount of time they waited before beginning to sing. When the stimulus stopped, the birds did not immediately switch back to their usual amount of space between song phrases. Instead, they gradually adjusted the space until they reached their usual speed of delivery. Todt and Hultsch concluded that the birds were truly interacting with the songs they heard, not responding in some automatic manner. The kind of song response seems to be a voluntary choice made by the bird. In a later study they concluded that the way nightingales choose to match song reveals a subtlety not seen in the song sparrow; rapid matching is meant as a kind of keep-away message with intent to warn, while matching after a break of a second or more is a kind of sonic greeting: Hello, here I am, I know that song too.

Each nightingale sings a series of phrases one after another in preferred patterns, much like the wood pewee but with fifty to a hundred different phrase types instead of three. Todt and Hultsch called these recurring groups of phrases *packages:*

> To explain our results on package formation in nightingales, we postulate two kinds of processes: 1. *A parsing process*. We assume that nightingales possess a gating mechanism that passes only a limited number of successively heard song types, and so generates unit-related segmentation of a long sequence of learning stimuli. 2. *A storing process.* We assume that nightingales possess several submemories, each of which can be supplied with data provided "package-wise" by the gating mechanism. These submemories process the received information in parallel and in a way which explains (1) the sequential association observed among song types in the package, and (2) the development of novel song types observed as recombinations of acoustic material stemming from song types in the same package.

This fascinating passage attempts to explain how it is that the bird can listen, learn, decide how to structure a series of phrases it learns, then recall it from memory years after the fact. A musician might call these packages *progressions* or the song *form:* this would suggest a definite level of musical intelligence in the bird, making it sound less like a computer program and more like a musical being.

In the wild, nightingales are thought to learn mostly from their elders, who spend much time feeding the young birds. In Hultsch and Todt's laboratory, birds would not learn songs from playback tapes alone, but required some kind of live model. A human caretaker would do, if the person started feeding the birds when they were young, say, around six days after birth. The most sensitive period of song learning is between two weeks and three months, although nightingales continue to learn throughout the first year, and also refine their repertoires later in life. Imitation seems much more important than improvisation with this species. Each bird learned the 214 master song types off a tape, but only if their caretaker or surrogate tutor was present.

Todt and Hultsch were amazed to discover that young nightingales could hear a song only ten to twenty times and then be able to perfectly reproduce what they had heard. No bird sang a song that it did not hear in the proper tutoring setup. Sometimes they acquired their packages by hearing them on the tutor tape, and deciding that a group of three or four phrases ought to stay together and in order, and sometimes they formed their own packages, which they favored in "song delivery." Then the packages of three to five songs were grouped into "subrepertoires" of three to five packages.

Birds reared together in the laboratory sometimes had an awareness that they shared each other's packages, a bit like the song sparrow matching activity. Hultsch had found the same behavior in wild nightingales a few years earlier. These layers of hierarchies, imposed on one another, resemble the cycles upon cycles imagined by Craig. This behavior suggests some limits in the nightingales' memory: we humans also "chunk" information to recall it more easily. Might the patterns also fulfill a musical purpose?

Since Sotavalta, no nightingale researcher has thought it a worthy subject to analyze the detailed structure of each song type or phrase. Hultsch and Todt do suggest that each motif might finish in such a manner that demands a reply. Maybe the final rattle is like a question mark? If you hear it as music, it's a sound that needs to resolve rather than a final cadence. When the Afghan musicians jam along with Baily's nightingale tape, it sounds like a call and response session that could go on and on for many rounds, with no resolution or release. Nightingales have called to one another for millions of years. Their bouts have no real beginning or end, voicing tendencies that might leave a trace of the bird's own evolution.

The most cutting-edge nightingale science is still at the level of stimulus and response. It's just gotten a bit more specific. A 2002 study by Hultsch and Todt's students, Marc Naguib and Roger Mundry, showed that nightingales respond most intently to whistle songs given in playback. They tend to match the whistles back, often at the same pitch. After playback stops, males respond with a whole slew of whistles. "These findings suggest that whistle songs have a specific signal value and that nightingales treat them as a special song category." What is special about these whistles? We can't say. Nightingale science is quite precise at articulating just how much we still do not know.

Can we be any more certain that nightingales are making music if the song brings pleasure to our ears? "The supreme notes of the nightingale envelop and surround us," wrote Lord Grey of Fallodon in the 1920s. "It is as if we were included and embraced in pervading sound." Yet he is not a complete fan. The song "arrests attention, and compels admiration; it has onset and impact; but it is fitful, broken, and restless. It is a song to listen to, *but not to live with*."

We long for similarities between us and the birds to make us feel more at home in their world. Perhaps animals' perception is farther from our own than we would admit. Sixty years ago Tinbergen noticed a stickleback fish aggressively displaying toward the window of his fish tank. What did he see there? Certainly no red-bellied fish that would indicate the traditional attack posture. No, the fish was striking toward a red mailman's truck far in the distance. Why bring in this story? Nick Thompson, the brown thrasher man, mentions it in his critique of anthropomorphism in ethology, saying that this tale shows that the stickleback has one strange way of reacting to the world. We should not imagine that we share much aesthetic sense with a fish! He really didn't like that truck.

Each animal species lives in its own ethological world. Aesthetics, should we believe they exist in animals, must be part of that. The starling never sings "-nee River." Song sparrows find matching songs to be a sign of aggression. Wood pewees' elegant songs are theirs alone. Why even claim to appreciate bird music for some kind of elusive, eternal essence?

Each living species is unique, but we are still all bound by the same cycles. Birth, experience, love, mating, travel, death. Each one of these phases can be expressed! Raw emotion leads to bird song and also to human art of all kinds. Something needs to be released, and what comes out is often wonderful. Communication and miscommunication both result from listening and playing along. Consider Oscar Wilde's story "The Nightingale and the Rose," where he turns that Persian nightingale tale upside down to imagine a bird trying to interpret human sentiment and performance and getting it all wrong.

A young philosophy student is desperate for a girl who says she will only dance with him if he finds a red rose. But there is none in the garden to be plucked. A nightingale in her nearby nest hears his plight. "Here indeed is the true lover," says the nightingale. "What I sing of, he suffers: what is joy to me, to him is pain." At once the difference between birds and men arises. We suffer in love while the nightingale just enjoys it! (Wilde's singer is a *she,* not a *he,* but literature never exactly matches life.)

There is only one way the nightingale can get the boy a rose—that terrible travail of Persian myth. A tree tells her the method: "If you want a red rose you must build it out of music by moonlight, and stain it with your own heart's blood. You must sing to me with your breast against a thorn." The thorn will pierce the bird and she will bleed into the tree, and a red rose will grow by morning. So love for the bird will strike from joy into pain and then death.

But she's ready to do it, and cries to the student with a song he cannot understand: Be happy, she sings, you will get your rose. "All I ask of you in return is that you will be a true lover, for Love is wiser than Philosophy." The student looks up, not comprehending, and only whispers, "Sing me one last song. I shall feel very lonely when you are gone." And then remarkably, he starts to *analyze* the music he hears: "She has form—that cannot be denied to her; but has she got feeling? I am afraid not. In fact, she is like most artists; she is all style, without any sincerity." If he only knew why she has begun to sing, and where it will end! All for him! "She thinks merely of music, and everybody knows that the arts are selfish. Still, it must be admitted that she has some beautiful notes in her voice. What a pity it is that they do not mean anything, or do any practical good." The boy remains a philosopher, trained better as a critic than anything else.

He goes to bed to dream of love, not listening closely enough to the bird to grasp what she was doing for him. In the morning the nightingale lies on the ground, dead, but on the very top of the tree stands a magnificent red rose, "petal following petal, as song followed song."

What luck, cries the student, and plucks the great flower. He takes the proud flower to his girl, but she just sloughs it off. It won't go with her dress, and another boy has already bought her some gemstones. "Everybody knows jewels cost far more than flowers." The student tosses the rose into the street, and a cart runs over it. "What a silly thing Love is," he decides. "It is not half as useful as Logic." It always makes us believe things that are not true.

The nightingale spilled all of her blood to use song to make a flower, which no one cares for after it fails. The bird and the human never understand one another. That beautiful suicidal music changes nothing at all.

The basic criticism of the romantics' love of nature is that they listened to birds and heard only themselves. If we are sad, the nightingale sings a sad song, and if we are happy, the same music is all about joy. Wilde reverses this "pathetic fallacy" and has the nightingale suffering because she imagines the young boy is consumed with passion, while in fact he is a lover of logic more than anything else. He, in a similar blunder, hears design in the bird's fatal song but no great wonder or force. He wants the flower but hears no connection between blossom and bird. Because the splendid rose gets him nothing in the end, he throws it out and goes back to his books, having learned nothing of love, nature, or life.

What will it take for us to learn what comprises the song world of birds? We need reason, passion, and diligence. Here are a few people who have taken time and effort to decode glimmers of meaning out of the surges and patterns of the sounds of birds. They have listened and waited, imagined and described. Music, science, poetry, practice, and theory intensify our awareness of nature's music without reducing the lingering wonder. If all the information doesn't bog you down, you may emerge from all the details with more ability to pay attention when you hear a bird sing.

It is a small step from playing a bird back his own song to playing him ours instead. In the 1920s, the British cellist Beatrice Harrison moved to the Surrey countryside and began practicing outdoors in spring. Nightingales began to join along with her, and she heard them matching her arpeggios with carefully timed trills. After getting used to her they would burst into song whenever she began to play. In 1924 she managed to convince Lord Reith, director general of the BBC, that a performance of cello together with wild nightingales in her garden would be the perfect subject for the world's first outdoor radio broadcast. Reith was initially hesitant: Surely this would be too frivolous a use of our latest technology? What if the birds refuse to cooperate when we're all set to go?

It took two truckloads of equipment and a bevy of engineers a whole day to set up what could today be arranged in minutes. The microphone was placed close to the nightingale's usual singing post. Harrison dressed in finery as if for a London premiere, though she sat with her cello in a muddy ditch next to the bird's bush so that the one microphone could pick up both of them. She started with "Danny Boy" and parts of Elgar's cello concerto, which had been written especially for her. No sound came from the bird. Donkeys honked in the distance, rabbits chewed at the cables, but no bird could be heard. This went on for more than an hour. Things didn't look promising.

Suddenly, just after 10:45 P.M., fifteen minutes before the broadcast was set to end, the nightingale began to sing, along with Dvorak's "Songs My Mother Taught Me." If Hultsch and Todt were listening, they would definitely hear song overlapping here. Was the bird trying to "jam" the cello message? Most of us would hear something more mutual—a mixture of bird and Beatrice, an attempt to blend. Doth the pathetic fallacy rear its ugly head—naïve anthropomorphism, or some moonstruck wish to hear music where there is nothing but practical noise?

I doubt many of the more than one million listeners who tuned into this broadcast were so skeptical. Never before had a bird's song or any other sound from the wild been sent out over the airwaves. The program was heard as far away as Paris, Barcelona, and Budapest, and many who had read the famous nightingale tales now heard one on radio for the first time. Harrison received fifty thousand letters of appreciation. After this late-night triumph she became one of the most sought-after cellists of her time.

The cello-nightingale duet was repeated live each year on the BBC for twelve years, and afterward the birds alone were broadcast until 1942, when the recording engineer making the show heard a strange, unmistakable droning sound that turned out to be the beginnings of the "Thousand Bomber" raid heading via Dover to Mannheim. He quickly shut off the sound, having the sense not to broadcast it during wartime. The recording was preserved, and you can hear it today, this strange soundscape of menacing bombers and incessant nightingales, singing as they always do, even in the midst of human destruction and the violence that comes with civilization. Even airplanes could not silence the nightingale. Here is a bird who cares nothing for the whims of men or the great noises we produce. Does he know his place extends far beyond the disasters of history?

After the war, science moved away from such musical analogies and interactions to focus on how birds are able to learn such intricate sounds. None of our primate relatives manage anything like this feat at all. In addition to a love of music, we share vocal learning only with a small number of bird species and some dolphins and whales. Birds are smaller, more common, and easier to cut up. In recent decades science has turned away from descriptive structure to peer directly into bird brains. It has found something more astonishing and revolutionary than anything we have seen or heard thus far.

CHAPTER 7

The Canary's New Brain

For three decades now bird song science has been moving from listening toward dissecting. Among songbirds, two species have most extensively had their brains and genetic material prodded and decoded. Years of detailed laboratory work has been done on the zebra finch, a closed-end learner originally from Australia, and the canary, an open-ended learner, popular as a pet for centuries, having been selectively bred for color and song by bird fanciers from a green-brown finch native to the Canary Islands.

The story of bird brain research is rich and complex, and it is most important because it has led to an astonishing result that no one anticipated: When an adult canary learns a new song, it grows new brain cells in the upper parts of its brain. Sometime in your life you have probably heard the warning: Watch how much you drink because you'll end up killing those precious brain cells and they won't come back. Once grown up we have all the brain cells we'll ever get, and from age eighteen on they just drop off one by one until we no longer remember why we're here. Through research on birds, this century-old dogma of brain science is being overturned.

Building on the discovery of adult neurogenesis in canaries, we might one day figure out how to grow new human brain cells at will, but that is still years away. From work on zebra finches, and the comparison of their song-learning systems with those of parrots and hummingbirds, we've learned there are some features common to all vocal-learning brains. In these aspects of brain structure, humans are closer to songbirds than to chimpanzees. Although gibbons sing elaborate duets in their trees at dawn, they are born with this ability, they do not learn it. Birds and humans share the ability to learn to sing, something no ape can do.

Until the early 1970s most bird song scientists were content to listen, print out sonograms, and play birds their own songs. Then they began to consider questions that required more intervention. For example: Does a bird need to hear its own song in order to sing it? Common sense would suggest that they do, but an assumption is nothing to science unless it is tested. One of Peter Marler's first students in his Berkeley laboratory, Masakazu Konishi, deafened young juncos, robins, black-headed grosbeaks, and white-crowned sparrows in their first weeks alive. Marler remembers how he did it: "When he was a beginning student, he wanted to try some surgery but he didn't have the right tools. So he took a lightbulb, broke it, and removed some of the wires inside to create a blade and a hook to pull out the cochlea of a bird from the inner ear." What did he find out from these birds who couldn't hear before they learned to sing? All of them, of all species, developed only raspy, buzzing, buglike sounds instead of their normally melodious songs.

You might squirm when you learn how this experiment was conducted. Many scientists doing this work believe the birds feel little pain when treated thus, while others feel that the techniques of their brain-prodding is as humane as it can be. All scientists who conduct this kind of research believe the results of their work justify the means. Konishi discovered that the effects of deafening were much stronger on young birds who had not yet firmly learned their songs. Deafening had little effect on adult birds who had already learned how to sing. Like Beethoven, they knew what to do and did not need to hear to keep at it. Konishi wrote that "the motor pattern of song, once stabilized, can be maintained without any sensory monitoring." Might the bird's brain be acting by rote because its wiring is set and established, even if he can no longer keep track of what he is singing?

No one dares do this kind of experiment on humans. What we know about human brain function began with observations on brains damaged in accidents. The notion that different parts of a brain control different functions goes back to one specific case studied by the French physician Paul Broca in 1861. One of his patients suffered a stroke, and afterward could only speak one word, *tic,* and was otherwise severely impaired. After the patient's death, an autopsy was performed, revealing a severe brain lesion on the left side of the frontal lobe, now called *Broca's area*. From this and other confirming experiments science learned that language within the human brain is localized, not dispersed. One specific region on the left side of the human brain controls vocal capacity. Animals without vocal learning abilities have no specialized vocal regions. What about birds?

In the 1970s Peter Marler moved to Rockefeller University in New York, and Fernando Nottebohm, previously his student at Berkeley, followed him as a junior professor. A native of Argentina, Nottebohm has a deep affection for songbirds and a love for their songs, but at the same time he had what Marler describes as enough "courage" to do the experiments that needed to be done to transform the study of bird behavior—which is part of ethology—into a fruitful branch of neuroscience where the inner workings of the brain as organ could be investigated.

Konishi showed that the adult bird who has already learned to sing can still do it well enough if he can no longer hear. The nerves that convey instructions from the brain to the syrinx, the bird's two-sided sound producing organ, must also convey some information back to the brain to ensure that the song is being sung correctly. In 1972 Nottebohm severed the left side nerves—no more song. He took another bird and sliced the nerves on the right—the song was sung just fine. Surprise? The nerves on each side did not function the same way. The left side of the syrinx seemed more significant for the song. Did this asymmetry extend back into the bird's brain? Until this moment it had been assumed that only humans had this kind of neurological asymmetry, with one side more important than another for different vocal functions.

At least two birds had to be impaired to prove that assumption wrong. Why do we get squeamish when we hear that songbirds are silenced in the name of science? Is it right to kill them at the height of their performance so we can examine the insides of their brains moments after they sing? "He sings his head off, and his head comes off," says leading next-generation bird song researcher Erich Jarvis, who has come to terms with this part of his work:

> When I first started as a scientist I worked with bacteria, I thought nothing of how many organisms we had to kill. When you get to a vertebrate animal like these songbirds, I did have concerns—what am I doing this for? My conclusion is that I am killing them to gain knowledge to help us understand how the world works and to help preserve the world. . . . For human beings, *knowledge is food.* We need to know as much as we need to eat.

He points out that most people do eat meat, and millions more animals are killed for food than for science, and they are generally treated far worse than laboratory birds with little public protest.

Jarvis's lab kills about two hundred birds a year, keeping an equal number alive for noninvasive behavioral experiments. "We are not cruel to these animals, we do try to kill them in as painless a way as possible." Killing has been necessary to discover most of what we have learned of the inside of birds' brains. This is part of neurobiological research all over the world, yet no clear, ethical defense of animal experimentation is agreed upon by the practitioners of neuroscience. They generally believe that the brain is a kind of organic machine, something we will never understand without taking it apart piece by piece. Only by such dismantling will we be able to fix it when it is broken.

This work has been fruitful, but scientists ought to spend a little time, as Jarvis has, trying to justify the practice rather than simply accepting its validity. Not even one bird should be slaughtered without some pang of guilt. Another bird song neuroscientist, who asked not to be named, agrees. "It's not good to kill animals without some remorse." Some scientists are afraid to speak out on this matter because they are worried they will become the targets of extremist activists, who might make their lives miserable with threats and harassment. "It's better," another tells me, "to stay off the public radar on this issue." If that is true then it is indeed unfortunate—we end up lacking a constructive dialogue on the legitimacy of killing birds to figure out how they sing. At least scientists should be encouraged to honestly discuss this part of their work.

Perhaps this is why many scientists prefer the term *sacrifice* to *kill* in describing the moment the bird's brain is severed from the body so analysis can begin. They realize animals lose their lives and are consumed as human knowledge. As amazing as their brains turn out to be, the birds do not sign consent forms authorizing us to take their lives for study. I hope that the ethics committees that are supposed to approve all animal experiments consider such issues. Some of you feel such treatment of living, breathing individuals is unacceptable; others may believe such sacrifice is justified because of the kind of progress this kind of science has made. Consider some of the results of Nottebohm's research: Canary brains, like human brains, are left-side dominant. Male canary brains have a much more developed song system than female brains, directly challenging the previous dogma that there are no significant differences between male and female brains in higher vertebrates. And Nottebohm's work is an essential step toward the possible application of adult neurogenesis through brain repair using stem cells.

Today there are nearly one hundred laboratories worldwide investigating the brains of birds. The more we learn, the more parallels to the human vocal learning system we discover. We know more about vocal learning pathways in bird brains than we do in humans, especially in canaries and zebra finches. Neuroethology is tremendously time consuming, and each species has very specific quirks and genetic differences, so the fewer species science works on, the better, and the easier for the knowledge to be collective and coherent, with each researcher's work building on the rest. Why these two species? Because zebra finches learn their complete song within ninety days of birth, while canaries are able to learn new songs well into adulthood, adding new sounds every spring. It was hoped that this difference might extend to a detectable difference in the two species' brains. And that's how the creation of new neurons in the adult brain came to be discovered.

Neurons are the messenger cells of the brain, using electrical and chemical signals to transmit information between different parts of the brain. Each neuron has a nucleus, and two kinds of extensions: treelike dendrites and a long stem called an *axon*. Other kinds of brain cells, oligodenrocytes and astrocytes, attach to the axon. Different sections of the brain control particular functions, and routes have been tentatively sketched between them when traces of neuronal activity are found to follow certain paths.

How are the connections discovered? Two basic ways. Tracks that link one region of the brain to another can be traced by injecting fluorescent dye into the brain region you're interested in, and over several days the dye spreads selectively from one region to another, through thousands of firing neurons. The bird is killed, the brain is dissected, and each section of the brain is meticulously analyzed and its color compared. Track tracing is the way to discover the overall circuitry of the brain.

It is also possible to record the electrical activity in two or more neurons simultaneously, and the connecting circuitry can be deduced based on the relation between the times each neuron fires. Precise mathematical comparisons can confirm the specific circuitry, but this method is hard to use to describe the overall system. You have to trace the tracks first to surmise which areas of neurons are likely to be connected to one another.

The same kind of electrodes that can damage a bird's brain in the name of science can also be used to monitor brain activity, now that the electronics are much more sensitive. Neurobiological study of the song system only blossomed because we learned how to record the bird as it lived and sang. The animals are awake—a far better way to study behavior, but not enough in itself. We can only trace behavioral firing in neurons once we know where to look, and that took a lot of dissection. Today MRI scanning machines enable science to study language areas in humans, but it is extremely difficult to keep a bird still long enough to use such machines on them. Resolution is also a problem: an MRI can barely distinguish the left side of the bird brain from the right unless you use an extra-strength magnet and special coils. Even then you only record blood flow, which may or may not be related to neuronal firing. And no bird will sing when its head is clamped still inside a machine.

Another recent technique of monitoring the bird's brain while singing is chemical, based on the measurement of gene activity at the molecular level inside individual cells in the bird's brain. DNA is inside all cells, and genes are stretches of DNA. Genes direct the manufacture of specific proteins. This process is called *expression.* When a gene is expressed, the DNA that makes up the gene is copied to make messenger RNA. The messenger RNA is then used as a template to direct the synthesis of a protein. Some genes, like the ones directing the production of human hair, are constantly expressed in the hair-making cells on the head, and, consequently, the proteins for producing hair are constantly manufactured in those cells. But other genes are expressed only in specific cell types, or only after a specific stimulus. Recently a gene has been identified whose protein products are synthesized only when the bird is singing.

How is this found out? Just after the bird is intensely engaged in song, the brain is removed, and frozen in dry ice. Then you place it on a cryostat, a machine that cuts extremely thin brain sections, ten microns thick. Claudio Mello and David Clayton discovered that the messenger RNA and protein products of this specific gene, which they code-named ZENK, is expressed whenever a bird hears or sings his particular song. The more he hears the song, the more protein products of the gene appear, but only in those particular parts of the avian brain involved with producing and learning how to produce the song. Says Jarvis, "Perhaps every time the bird sings, ZENK is induced to synthesize and replace proteins that get used up during the act of singing." Half an hour later, when the zebra finch wants to sing again, the proteins are back and the pathway is ready.

Using all these methods together, the areas of the bird's brain active in singing have been mapped out, together with different pathways that function at different times. The map that emerges is full of enigmatic places and ever more sure arrows, refined by hundreds of experimenters working all over the world for nearly thirty years. The whole thing is called the *song system*—how far away it all seems from the melodious and essential songs we have been musing about in previous pages. The following is a simplified description.

One key section of the forebrain, if severed, causes birds to lose their song completely. This is the region most crucial to song production, where the most neural activity appears as the song is sung. This region of the upper brain was first called *hyperstriatum ventrale pars caudale* (HVC), and later named the High Vocal Center. Neurons from the HVC project to a second region, further forward in the bird's brain, called the Robust Nucleus of the Arcopallium (RA) because of its clear and robust appearance on stained sections of brain tissue. Other neurons project to a different area, whose function was at first less clear because lesions there did not lead to loss of song. This the scientists named *Area X.* Later it was discovered that the pathways through Area X are most active when birds learn new songs, not when they perform tunes they already know.

These are the three most important regions of the song system: HVC, RA, and Area X. The more research is done, the more intricate the range of pathways becomes. Some areas have recently been renamed to emphasize their similarity to the human brain. Isolate singing from everything else the bird does, and this is where it goes on in the brain. Describe the movement along these pathways and it starts to sound like instructions out of a technical manual for fixing

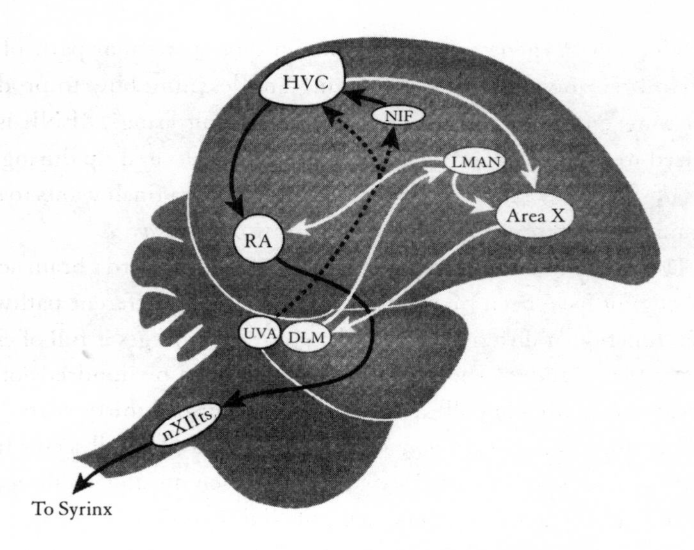

THE MAJOR NEURAL PATHWAYS OF
THE SONG SYSTEM IN THE BRAINS OF BIRDS

The named areas of the brain are loosely differentiated regions, not clearly definable organs. The pathway in the front of the brain from HVC to RA is followed when the song is sung. The rear pathway through Area X, DLM, and LMAN is only used when the song is learned. The other pathways are involved with breathing and sound production. New details are being discovered all the time.

some very abstract kind of machine. Seeing song like this seems far removed from experience, either ours or the birds'. The brain does not come easily to the design metaphor—is it really a machine made of meat? These areas of the brain are identifiable to the naked eye by the boundaries between cell types, but they are not discrete like the intricate gears of a Swiss watch. Our understanding of the song system thus far is literally a sketch of terms and lines over the reality of a mysterious gray matter, folded or smooth, like or unlike.

The specific simplicity of zebra finch song makes it ideal for studying exactly how and when learning takes place in the brain. The

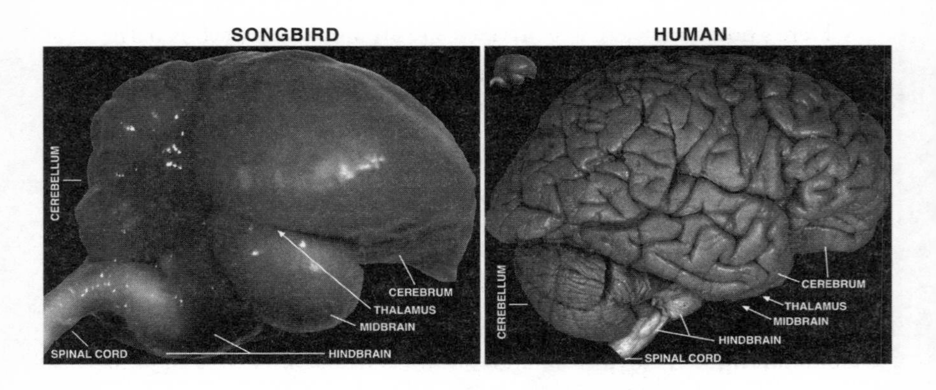

A COMPARISON OF BRAINS

Note the relative size of the bird's brain next to the human brain in the top left of the right image.

greater flexibility of canary song (although still nothing like the super-complex songs of the nightingale or marsh warbler) makes it ideal to investigate whether their adult brains are uniquely flexible.

Nottebohm and his collaborators had been intrigued by the strong differences between the brains of male and female canaries, especially during the spring mating season. The HVC and RA in male canaries is about twice as large in spring as in fall, and this was determined to be linked to the presence of the hormone testosterone. They had already determined that if you injected a female canary with testosterone, it would start to sing like a male. A testosterone-drugged female would also grow neurons in the HVC with long axons and many secondary dendrite branches, the kind of neurons males naturally develop during springtime, when their hormones are raging. So hormones had to have something to do with why birds sing. The brain structure seemed so fluid and flexible—was it really all a rearrangement of the same set of neurons, as Nobel laureate Ramon y Cajal had defined in 1913? According to his theory, "Nerve paths are something fixed, ended, immutable. Everything may die, nothing new may be rejuvenated." We now know this is wrong.

Here's how they were able to track the moment a new cell is born: A cell that is just about to divide synthesizes new DNA. If you inject an animal with a radioactive form of thymidine, a substance that collects just before DNA synsthesis, the thymidine will concentrate inside the nuclei of cells that are about to replicate. (Thymidine becomes the T in the familiar DNA nucleotides GCTAGGTCA.) The chromosomes themselves will be labeled. When the cell so marked does divide, half the radioactive DNA will be in the nucleus of each of the two new cells, so they will be marked as well. Nottebohm and Goldman injected thymidine into adult canaries for several days in spring, then stopped the injections and let the birds sing for one month. To their amazement, they found that as many as one percent of the neurons in the HVC were labeled for each day the birds received injections, revealing that the original cells had been dividing. As they investigated further, it appeared the new cells had originated in another part of the brain and migrated into the HVC, where they turned into neurons as they connected themselves to existing neural circuits.

This was the beginning of a revolution in our understanding of brains. Once the process of new cell generation in the adult canary's brain was confirmed, researchers looked again at other creatures and discovered that the same thing was happening in mice, and in primates like marmosets and macaques. Neurobiologists who had spent decades cutting up mouse brains wondered how they had missed this. Did we really need to scare all those teenagers with tales of dying brain cells? New neurons in the mouse hippocampus were enhanced by learning and reduced by anxiety. Jarvis believes that one reason adult neurogenesis was overlooked may be that so many laboratory animals live under such severe stress.

Happily singing, learning canaries were the first species to reveal neural replacement. If we could figure out how to stimulate the growth of new neurons in the human brain, we might be able to direct this knowledge toward the repair of brain injuries such as the stroke that led Broca to first realize that brains had distinct areas controlling specific functions. Starting with humble singing canaries, brain science is heading into a new paradigm, a vision of neural plasticity, where the cells in the brain are changing, not set or fixed. This is the canary's wonderful secret.

Why are only *some* neurons replaced in the adult brain? It is now understood that the new neurons are temporary, and that they replace older ones that have died. Nottebohm believes that the "spontaneous replacement of neurons calls for a new theory of long-term memory." Fresh neurons are added to many parts of the forebrain of adult canaries and zebra finches, but rarely in other parts of the brain. New neurons are added every month of the year, but the total number of HVC neurons remains constant in canaries even when the size of the HVC grows during springtime mating season, because older neurons die off. The growth of new neurons definitely has something to do with the singing of new songs. If the birds no longer sing, the total number does decrease.

Why would a healthy brain want to replace neurons that are functioning perfectly well? The death and birth of neurons should tell us something about the limits of a brain's ability to learn. Consider a cell working as part of the memorization process of a song. For these neurons, "the acquisition of long-term memory may be a onetime thing," according to Nottebohm. Perhaps the neuron cannot learn again once it has done its job. Perpetual replacement of old neurons by new ones may be a method for the spontaneous rejuvenation of key brain circuits. Under this scenario the brain is constantly remaking itself, a far cry from Cajal's vision of a well-formed brain gradually wearing down. This may explain why, with virtuoso species like the lyrebird and mockingbird, the older the singer, the more developed the song.

Neuronal replacement has revolutionized our grasp of the anatomical actions at work when bird song is produced. Nottebohm is pleased by the value of his discovery, although he says "it used to be much more fun when nobody believed it. In science, by the time everybody tells you it's true you have to scratch your head and look for another business." In spring 2004 Brandeis University awarded the Lewis Rosenstiel Prize for Basic Medical Research to Masakazu Konishi, Peter Marler, and Fernando Nottebohm for their pioneering work in the ethology and neurology of bird song. E. O. Wilson presented the award.

Connecting what we have learned about neurogenesis in canaries to repairing damaged human brains will be a long journey. Most of the birds studied thus far have lived in very controlled circumstances, under strict and stressful confinement. Nottebohm, as methodical as ever, urges us to remain cautious in our claims. "It would be wise to look more closely at adult neuronal replacement in a diversity of free-ranging animals leading a normal life, because I believe this material furnishes the best examples of what nature can do. . . . Before we try our wizardry, we should find out how nature uses it."

The pathways inside birds' brains have been charted as they learn new songs and perform polished ones. But the link between this knowledge and what goes on in the human brain is largely unexplored. Erich Jarvis says that this neglect stems from the belief that "humans are much more special than we really are," so people are reluctant to accept serious parallels between these tiny smooth bird brains and our own, with its heavily folded outer cortex that is thought to separate humanity from the rest of life. Science, Jarvis notes, recognizes that human vocal communication includes three parts: speech, which is the production of sound; language, which is the sequencing of sounds into meaningful structures; and comprehension, which is the perception and understanding of language. All three phases are necessary to get the message across.

When it comes to birds, "there is no equivalent distinction between language-like and speech-like properties. One word, 'song,' represents both production of learned sounds and their sequencing." Information is not the most salient quality encoded in bird song. There is form and purpose in birds' sounds, but no message to be extracted from the sound that makes any sense without the sound's original shape. In this way song is more like human music than human language.

I'm a bit surprised to hear Jarvis agree with me. "The song many birds sing is more akin to music than language because when you study sonograms in detail, the syllables change gradually from one form to another. They are not usually as distinct as the sounds of language." He also feels that most birds sing more for mate attraction than territorial defense, with melodic sounds more attractive and noisy sounds more territorial, a fact some of the studies we have considered in other chapters confirm. His own work on hummingbirds has shown that these tiny birds with very high-frequency voices actually combine short, innate calls into learned songs of deep structure and complexity.

The larger context of Jarvis's work is the evolutionary place of vocal learning—why do so *few* kinds of animals have it? Humans, whales and dolphins, and just a few orders of birds. There must be a stronger form of natural selection *against* vocal learning, perhaps the threat from predators that such time-consuming vocalizing must imply. The faculty of learning must have evolved in very specific circumstances. We don't really know what they were. And what about the purpose of music in humans? "Think about Jennifer Lopez, Ricky Martin," says Jarvis. "Of course we use music to attract mates. Just because it's more complex in humans doesn't mean it's not the same kind of thing."

And just because it's simpler in birds doesn't mean that they might not also sing for the sake of singing itself, for the beauty that music and music alone can express, whatever genes are also expressing themselves in the process, and whatever pathways of neuronal electricity have been identified. What has thus far been discovered is intricate and complex, but it is still far from giving us any grasp of what happens when the specific details of the song itself are produced and processed. Our understanding of the brain, however it advances, is still far from the notion of how the *mind* operates when making sense of sound, either in humans or in birds.

Nottebohm charted how song as a whole passes through the brain, but what of the specific features of more complex songs? Martine

Hausberger has examined the auditory perception of starlings at the cellular level, to see how neurons in different parts of their brains react to very specific whistle sounds. Remember that starlings sing two kinds of whistle sounds. One downward sweep kind is sung by all male starlings; this one she presumes allows for recognition of species, and specific populations. The second class of whistles are more diverse, and they might be peculiar to each individual bird or shared by a small group of birds living together. To which sounds do specific neurons in particular parts of the starlings brain respond?

Hausberger caught six wild adult male starlings and surgically implanted a ring-shaped steel chamber on the top left of their skulls, while the birds were under anesthesia. After the surgery the birds got two days to rest. Then the animals, while awake, were wrapped in a wing cuff to prevent excess movement, and the neuroelectrical recording device was attached to a head holder. Fifty-nine different stimulus whistles were presented in various patterns. The songs were presented ten times. Three hundred and twenty individual neurons were tapped. After the final recording session the birds were sacrificed with an overdose of the sedative nembutal and their brains were frozen and sliced into layers fifty microns thin.

Only six birds killed and plenty of data. Out of the 320 neurons checked, 232 showed a significant change in spontaneous firing when a stimulus reached them. Twenty percent responded to pure tones, 80% responded to whistles. Certain neurons responded to whistles with certain inflections. In the HVC there was the greatest response to whistles similar to the bird's own song. "Such plasticity suggests that if there is a template to guide song learning, it is probably a program open to new species-specific information." What kind of information? We still don't know, but scientists continue to hope for a reason for the music. How can the starling maintain its species-specific music that includes an astonishing ability to imitate and improvise? Can we track the bird's ingenuity inside its brain?

We would never conduct this kind of experiment on a human being. Kill a virtuoso in the heat of musical passion, freeze his brain, and examine all the neuronal connections? Sure, it is easier to monitor larger, more cooperative, accepting human subjects than it is to keep tabs on birds, who are less aware of what is being done to them. Still, some of the most interesting insights on human musical ability and the brain come from subjects whose brains have been marred.

The composer Maurice Ravel suffered a stroke while swimming in 1933, at age fifty-eight. He was still able to enjoy music, but he could not create or perform it. Ideas were locked inside his brain; he felt they were there, but he could not do anything with them. He was a victim of aphasia, indicating that the stroke was on the left side of his brain, blocking his ability to put words or sounds together in a meaningful way. Yet he could still *feel* music, suggesting that the ability to create music happens on the same side of the brain that creates language, but the ability to listen to and love it is retained somewhere else. Does the appreciation of music use the right side of the brain while the making of music requires the left?

Such tales of the separation between musical and linguistic comprehension are often considered anecdotal by the strict standards of scientific rigor. Recent human brain studies claim there are more similarities than differences between how the human brain processes music and language. In both there are hierarchical relationships between syllables and sounds, and when the brain hears patterns of sounds that it is able to make sense of, either musically or linguistically, it responds in distinct regular patterns that can be measured over time. Aniruddh Patel and Evan Balaban believe we need better research on healthy normal human brains, not impaired virtuosos. The two San Diego neuroscientists attached 148 sensors to the scalp of a volunteer and tracked the electrical current through different parts of the brain as melodies and scales played. They then repeated the same experiment while inserting random bouts of noise.

The brain's activity turns out to be more evenly pulsed when it hears music than when it hears unrelated notes. It is more regular when it hears scales than more disjointed melodies. We do not know why. Patel and Balaban hope that using music to study brain-region response in human beings will add the element of duration to brain reaction studies, which previously took single snapshots of our brains and tried to deduce what was going on, much like birds struck down at the moment of ecstatic singing and their brains immediately frozen into a steady, sliced-up state. Cortical response unfolds over time as we listen to music being played. Our brains are engrossed in the sound, and the discernible results may be much clearer than when we're listening to talk, because music itself is often more rhythmic and regular than language, somehow closer to the deep computation at work in the firings and sensings of the brain.

Most animals do not possess the ability to string sounds together to convey specific, constructed messages. The closest thing to language that we have figured out in the animal world is the waggle dance of the honeybee, in which one bee tells others in the hive where the flowers are and perhaps where the colony should move. Karl von Frisch won a Nobel Prize in 1973 for this discovery and there is nothing remotely like it in the rest of the animal world. When whale songs have been analyzed for information content, not much has been found. Never mind, there is not much information in a Bach chorale either, but there is plenty of order, beauty, and even a touch of purpose. To me it is easier to find music than discourse in the animal world, probably because form and order in natural sound come long before syntax.

Scientists working in the highly speculative field of biomusicology have suggested that human ancestors may have found the pathway to developing our linguistic potential through the more basic activity of music making. Swedish biomusicologist Nils Wallin said that protohumans learned to control their vocal chords through the emulation of animal cries. Imitating their movements may have led to dance. Gibbon expert Björn Merker has said that human beings "are the dancing apes. Man must dance," and that's part of what led our culture to advance.

Groups of people moving in rhythm together bring ritual, and that allows us to synchronize our actions. We learn how to jam. People are connected. Society is built. Cognitive scientist William Benzon imagines that the music of human ancestors was "oriented toward other people and the sounds one creates with their help and cooperation, rather than toward arbitrary objects in the nonhuman world." What he calls "ritual musicking" was "practiced by hominids who had survived the long journey through the steppes, a march that required them to become virtuosi in the techniques of group safety." Benzon believes that the synchronization inherent in traveling group life later led to dance, ritual, music to listen to, and finally the more reflective faculty of language.

After several hundred years the "speculative music" of Kircher and Hawkins has returned as science. If specialized organisms such as birds have evolved the ability to make music, then it may be much more fundamental than we previously thought. I wouldn't call it simpler than language, just more primal. Time is the key. Over a blip of time a zebra finch sings a simple song, which he has taken two months since his birth to learn, according to a process that took millions of years to refine. Once he's learned it he'll always sing it the same way. That remarkable trajectory of song crytallization is one reason that their brains have been electrocuted, abused, frozen, sliced, and diced. This is one way to get at their workings, but not the only way. With its static approach, it doesn't take stock of time.

The most valuable thing about the zebra finch as a research subject is that they do all their song learning in a fixed interval of just a few months. But how exactly is the song learned? Does it happen haphazardly or in an organized way? Sure, the song is brief, quite exact once it is learned, and varying from individual to individual. We can listen to it, slow it down, chart it in sonograms, look at the units of design, following our own intuitions. But are those units the bird understands?

In the early 1990s Israeli biologist Ofer Tchernichovski came to upstate New York to work at Nottebohm's Rockefeller University laboratory, the woodland hideaway where much of bird song neuroscience has been pioneered for the last thirty years. At first he was entranced simply by the songs themselves. "When I first came to Millbrook I spent several hours each day just listening to the birds singing. Of course I couldn't *tell* my supervisors that's what I was doing, they wouldn't pay me for that. But I have always learned more from intuition than anything else—I wanted a feel for the songs."

Tchernichovski guessed that the birds learned their simple, specific songs in an organized process. For a scientist, intuition is not enough: he needed as much data as possible. He first needed to teach the birds to sing in very rigorous circumstances. Thirty baby zebra finches were hatched at roughly the same time. Both parents raised their chicks together until two weeks after they hatched. Then the father was removed (so that the babies were not attuned to his song) and the mother raised the chicks alone until thirty-five days after birth. Then each young male was placed alone in a cage. In each cage was one plastic model of a grown-up zebra finch. Inside this toy bird was a tiny speaker.

There were two keys in the training box. Pecking either key would trigger the playback of two identical repeats of a single complete song recorded from an adult bird. That's the only song these babies got to hear, and that's the song they gradually learned over the two-month sensitive learning period, beginning when they were one month old, ending when they turned three months. Every time they pecked the keys it would trigger the song, but only up to ten pecks per training session, one in the morning and one in the afternoon. After that no more sound would come—this was a highly controlled learning program.

With the immense digital recording capacities of today's computers, every single sound made by every baby zebra finch was recorded during the entire two months of the sensitive song-learning period. Each bird produced between one and two million distinct song syllables during this time, resulting in an immense amount of data. What to do with all this? Partha Mitra, a mathematician now at the Cold Spring Harbor Laboratory, devised a computer program to analyze this huge wealth of sound data. Let the computer crunch the data while the rest of us sleep and you'll end up with something no previous bird song study had amassed—more information to analyze than anyone had been able to previously collect. By this method Tchernichovski and Mitra hoped to avoid any human perceptual bias in making sense of the songs, thus approaching a kind of digital phenomenology far removed from that practiced by Nice, Craig, and Sotavalta.

The first step of the automated analysis was to determine which sounds are similar and which are different, using mathematics to identify the units of the song rather than the intuition that has guided nearly all previous attempts to analyze the structure of bird song. Tchernichovski believed that "we need to get beyond our own intuitions to figure out how the structure of the song makes sense to the birds. That information is essential to decipher the learning process." So every syllable produced by the birds was compared and analyzed using a simple set of features: duration and pitch are the easy ones, but zebra finch sounds are notoriously fast moving, noisy, and complex. The program additionally analyzed entropy: How random was the syllable? "Goodness" of pitch: How clearly is the pitch expressed? How much is the bird "in tune"? The collective result from examination of these features gave the researchers a mathematical way to accurately categorize and identify the appearance of every sound made by the birds.

The next step was for Mitra to develop an algorithm to cluster the syllables and track the evolution of these clusters over the whole two-month sensitive period. With all these parameters, Tchernichovski could display the data in various multidimensional animated graphs, which ended up looking like abstract video art, moving, pulsing through time. Whatever parameters he picks, a clear order appears on the screen—a dynamic order—that can be viewed in different ways. This is the most important technological advance I have seen in charting bird song since the sonogram. "I too," says Tchernichovski, "started with ugly sonograms. They have a high embedded dimensionality—time versus frequency. But in our approach we can change the dimensions at will, and watch them move!" These new dynamic tools can visualize the entire struggle of the song-learning period, charting the baby zebra finch's entire two-month education, a dynamic report card that presents a new picture of how a single song is learned.

Before this experiment scientists believed that zebra finch song development was not a highly structured process. "The first time the bird hears the tutor's song, he just seems to go into a daze," says former Millbrook researcher Jeff Cynx, now a professor at nearby Vassar College. "The development period has always been a black box. Ofer's really opened up the box, that's an astounding achievement." With their vast set of data, Tchernichovski and Mitra have shown that song learning is a patterned process. The birds learn a syllable one day, forget a little bit overnight, and catch up to where they were the next day. Song learning emerges and appears as a gradual process, but it is finally revealed how a slow course can lead to the abrupt learning of new syllables.

How does the bird get from ABABABAB to ABCABCABCABC, as a step along the way toward learning the whole song, which might be ABCD? Simple: the new syllable is gradually introduced. The graph of the learning process can be imagined in the style of Wallace Craig: ABABABABABABCABABABABABABABC, ABABAB ABABCABABABABABABCABABABABABCABABABC ABABCABCABABABABABCABABCABABABABCABABC ABCABABABCABCABCABCABCABC, this over a matter of hours and hours. He's getting there.

The first syllable learned is not always the first note in the song. Each multisyllable motif turns out to have some syllable that is the kernel, the core, the first one that is learned. It is usually one from the middle. "We don't know why they do this, but we are trying to understand how."

Tchernichovski doesn't think "Why do birds sing?" is a scientific question.

> Why does your heart beat? To pump the blood? Why does it need to do that—some design feature of the human animal. But I'm no complete Darwinian, selection doesn't explain everything. The actual be-

> havior of most systems is not explained by their functionality. No biological systems are optimal in their overall design. Nature has more than that, the world is computing all the time. There are a lot of emergent properties that just come with the territory.

Therein lies the rub. No scientist worth his data set really wants to tell you why birds sing. There's enough to examine in the quest to figure out *how* birds sing, and especially how complex behavior is learned in a very short time. The brain works exactly and methodically to get there. "Animals *are* machines," says Mitra, "but they are unlike any machines we are able to build or comprehend." I'm not so sure. Even though their behavior and education are found to follow exact rules, the actual life of birds lies far beyond the scope of mechanistic thinking.

Scientists, it turns out, squirm as much in their seats when I ask them "why do birds sing?" as when I ask, "are you sure it's all right to kill and torture birds in the search for the truth?" Neither line of inquiry sits comfortably within their line of work. Science has learned enough about the brains of birds to see that they have incredible abilities we hardly understand. Who feels comfortable chopping the head off an organism like that? Male songbirds need to defend territories and attract mates—isn't it remarkable that they use the same music to do both these tasks? Who would have expected something so beautiful out of evolution?

Ofer Tchernichovski knows that his pulsing patterns and triangles of moving plot points are beautiful. His intuition tells him to trust this beauty while his scientist's smiling confidence assures that it's backed up with millions of units of sung data. But the patterns don't explain why the song is there. Fernando Nottebohm summarizes the whole paradigm of observing birds learning and responding to songs—"You've got to go to each species and ask them what they like to hear." He warns me to stay the course. "This whole music approach of yours is really a *distraction*. What the birds like might not be pleasurable to our ears. You have to address the question of 'why?' to the birds themselves." So why, I ask him, is the result of what they sing so often beautiful to our ears, and if not obviously like our music, then built out of patterns, figures, order that is beautiful in a mathematical sense? This cannot only be coincidence.

Nottebohm turns to me and asks, "Well, why are *you* so interested in bird songs?" And I tell him the whole story, how I went to an aviary to play music with a bird, and the bird surprised me, changed my music, and set me on this quest, but as I go on, his attention seems to wander. "Ah," he looks out the window at the forest, hearing a few strange sounds therein, "the *why* question is such a bad question, isn't it? People ask you why you do something, and at that moment your mind confabulates, and you have to come up with a sudden answer."

Easier to imagine that the inside of a bird might tell you why he sings, but we humans know ourselves too well to trust our own quick answers to such ultimate questions. Too many variables for our introspection to be trustworthy! I continue to be charmed by the power of the most complex songs. I'm amazed by what brain dissection has revealed, but there is so much more to be learned from the song as it is sung. We need all possible tools to figure out what holds the music together, from beginning, through the middle, to the end.

THE MOCKINGBIRD

CHAPTER 8

Listen with the Mockingbird

EACH YEAR WE LEARN MORE about songbird sound production. After the brain, the syrinx is the most important organ in the production of sound. Its name comes from those mythical pipes of the Greek god Pan. The syrinx of most birds has two sides, each able to produce a separate sound simultaneously, which is one of the reasons birds can sometimes seem to be singing several songs at once.

Songbirds have five pairs of muscles surrounding the syrinx, and another pair above, on the trachea, that push indirectly on the resonating tubes. Lyrebirds manage astonishing vocal feats with only three pairs of muscles. As in the brain, the left and right sides of the syrinx do not operate symmetrically. The left side is dominant. The rest of the bird's vocal tract, from the trachea up to the beak, serves to modulate the pure, flutelike resonance of many bird sounds. The higher the pitch of the desired sound, the more open the beak needs to be. As registers and sides of the syrinx are switched, the bird opens and closes its beak with rapid movements. No one knows exactly how all these muscles are synchronized, but recent work on the mockingbird suggests that most singing birds share the same methods of sound production.

Duke University biologist Stephen Nowicki has compared a bird producing a complex song to a human musician playing a difficult score at a fast tempo. Imagine a bird producing a high repeating trill, *brlllllllllllll . . . :* Each fragment of the trill has a precise form, generated by a vibrating membrane in the syrinx, the shape of which is set by the tension of the surrounding muscles. Respiration from the lungs through the trachea produces the airflow necessary to keep the sound going. This involves a series of "minibreaths" through minuscule breaks in the sound. The beak opens and closes following the frequency contour of the notes. The bird may have to dance or assume a particular stance while the song is made. All these movements must be repeated precisely, one after another, over and over. More than brain alone, singing requires the synchronized work of every part of the bird.

Ayanna Alexander, a graduate student in Ofer Tchernichovski's lab at City College in New York, is trying to learn to play the zebra finch syrinx herself. "First I need to take it out of the body," she admits. "Any adult male will do, we have a lot of extras. I open up the chest cavity. I cut the trachea high, near to the beak, very close to the top, then I just lift it out." The wrenched-out syrinx must be quickly prepared.

> I cut a small piece of one of the muscles in the front, and pull it down to the point of where it's just connected to the third broncheal ring, where I attach a little piece of silver wire to the muscle slice. This is my first handle. Then in the back, I cut one of the dorsal muscles, and attach another wire. Both of these are connected to long wooden sticks, so I am able to move them with my own hands. I attach a vaccuum or a pump to the end of the syrinx. I turn on the air and move the handles around.

Voila! A human can now play a syrinx—one step closer to physically becoming a bird.

The disembodied syrinx will only stay moist for several hours, so Alexander doesn't have too much time to practice. So far she's only been able to manage some high thin whistles, a far cry from what the birds can do many times a second. She needs a new syrinx every day. "I feel really bad about killing these birds. We teach them songs, they sing, and then I kill them. It doesn't feel good. I cannot really justify it, I'd rather not kill." She hopes to soon move on to practicing on some kind of mechanical, artificial syrinx, built at a larger, closer-to-human scale. Then she will have more leisure to learn her instrument, and discover just a bit of what it feels like to sing *as* a bird sings, something no scientist, not even Wallace Craig, has previously considered seriously.

Even a single trill sounds like a Herculean task. Imagine what it takes for a virtuosic bird to blend all his world's bird sounds together in a deliberate composition, one that is never quite delivered exactly the same way. It's the song of the mockingbird.

In 1855 Septimus Winner, composer of several of the nineteenth century's most popular songs, wrote his biggest hit of all. He published most of these songs under the fictitious name of Alice Hawthorne, thereby becoming one of few men of that era to disguise himself as a woman to get ahead in the public realm. Winner sat in his living room listening to a caged mockingbird singing in his neighbor's yard. Suddenly the song became a duet when a second singer joined in. Another bird? Turned out to be a young barefoot boy whistling a tune with the mockingbird. The elements coalesced, Winner had his theme. Here's how it goes:

I'm dreaming now of Hally,
Sweet Hally, sweet Hally,
I'm dreaming now of Hally,
For the thought of her is one that never dies.
She's sleeping in the valley,
The valley, the valley,
She's sleeping in the valley,
And the mockingbird is singing where she lies.

Listen to the mockingbird,
Listen to the mockingbird,
The mockingbird is singing o'er her grave.
Listen to the mockingbird,
Listen to the mockingbird,
Still singing where the weeping willows wave. . . .

Upon publication, "Listen to the Mockingbird" spread faster and farther than any other song of its day. Abraham Lincoln said, "It is a real song, as sincere and sweet as the laughter of a little girl." People danced to the tune on the White House lawn when Lee surrendered to end the Civil War. King Edward VII in England remembered whistling the tune as a tyke. By the tail end of the nineteenth century it had sold more than twenty million copies in sheet music worldwide, making it one of the most popular songs ever. When you consider that in 1850 the whole population of the United States was only thirty million, the song's fame seems greater than any hit today.

What does the song actually glean from the real mockingbird melody? There is something of the rhythm, the reiteration, then a coda of contrast and respite. It's nowhere near as exciting as Walt Whitman's inhabitation of wild, repeating beats in "Out of the Cradle Endlessly Rocking," but you can hear in this song some homage to the American bird with the greatest musical stamina. They can go on for at least half an hour without stopping. Fred Lowery's mid-twentieth century recording of the song incorporates imitations of cardinal whistles. If any bird sounds like it is making organized intelligent music as it sings, the mockingbird is the one that does it at a speed humans can conceivably hear and decode.

Yet we have never taken this bird as seriously in the New World as the nightingale has been praised in the Old. Its song is more often called a mocking taunt than an outpouring of inscrutable love. Whitman did hear rhythms and tones that would drive American poetry vehemently onward (Blow! Blow! Blow! Soothe! Soothe! Soothe! Loud! Loud! Loud!), but most people imagine the mockingbird simply copies all the other bird sounds in his midst. If you listen closely, though, you will hear something much more remarkable: he composes his own precisely structured music out of the sounds around him. It is musically satisfying and beautiful, and no one has figured it out. The lines from the song sound familiar, but very few of us have taken the time and effort to listen to the mockingbird.

The real bird imitates all in his path, with clear and graspable rhythms. Evenly paced clicks. A break. The same thing sung higher and faster, faster, then a quip, a turnaround, a stop. Space. Another melody, a game played with that. Rules you think you almost catch—twists you don't expect, like a fine jazz solo. It's all alone, at the edge of the field. He's got his territory, he's looking for his mate. Then he finds her. Then he doesn't stop. He keeps singing. Singing on when there's no more need. If he's taunting anyone, it must be us—Fool, you think you can explain me! You think I tease you with my abilities? I sing the song of the world, the recombination of all that I hear. Listen in, and listen good. You only make fun of yourself if you think you can dismiss me or decode what I say. Sit at the table, endlessly mocking, but my song will endure far longer than your need to laugh it away.

Perhaps I am being too hard on us humans. What would it even mean to figure a song out? We can analyze Beethoven and come up with things to say about why it is great music, but that doesn't teach us how to write more Beethoven symphonies. You can try to explain what makes John Coltrane's saxophone solos superior, but there will never be another Coltrane. Music is both mathematical and momentary, following rules to break our hearts and souls.

Before there were scales or pitches, music emerged out of the cadence of life. Not just a single beat but the repetition and spacing of sounds enjoyed for their own sake. Rhythms must keep us enthralled and not bore us. One two three four five six. One two three four. One two. One two three four five six. Five. Four. Five. Four. How many times does the mocker repeat? He squeezes, switches, tests a swoop, then another, and finally a blue jay scream. He's compressing, transposing with the sound just the way my musical software plays with sound, morphing it from one thing into something else. Why does he do it? Why do I do it? To experiment, to play, to turn an inspiration into a riff I can stand behind. Here's a melodic question: *boo pe ah poo,* and again, *boo pe ah poo,* and then quickly, *sneep!*—An opposite sound, contrasting, to bring the phrase to a close. Six. Five. Six. Four. Four. *Sneeewwr. Sneeewr.*

The mockingbird song is long. Of birds we have considered thus far, only the marsh warbler approaches it in breadth and complexity. The warbler's song could be called more complex, because it streams out imitations of brief other-bird fragments at super-high speed. If *Acrocephalus palustris* is the Charlie Parker of bird soloists, *Mimus polyglottos* is closer to Ornette Coleman, repeating rough riffs, swinging with just the right space in between. Decoding the mockingbird seems much easier than the marsh warbler, because he swings closer to human time, in graspable units with a beat. The rhythm and diversity are well within our listening range—all the more surprise that no human has tried to unravel it.

The mockingbird doesn't simply repeat one imitation after another in the manner of a slowed-down marsh warbler. Not at all—he combines his repertoire of imitations into precise rhythmic patterns, groups of three to six, then combines these groups into what scientists have called "bouts." It is all a bit subjective, I must say. Listening as a musician, I hear each imitated song as a note grouped into motifs, which are then strung together into phrases, each with a definite musical feel. *Ch ch ch. Te ḳoo te ḳoo te ḳoo. Ka ḳa ḳa ḳa ḳa. Whrsneep.* At this level the structure sounds clearly musical. Whether each five- to fifty-second phrase is actually a complete song or merely part of a larger song that takes many minutes to deliver depends mostly on how you choose to listen to the bird or imagine that the bird listens to itself.

Peter Marler thinks it's mostly a question of how much space there is between phrases, but since the mockingbird's phrases are so regularly spaced, it is hard to know what the correct unit of analysis should be. At the risk of being totally unscientific, let's examine in

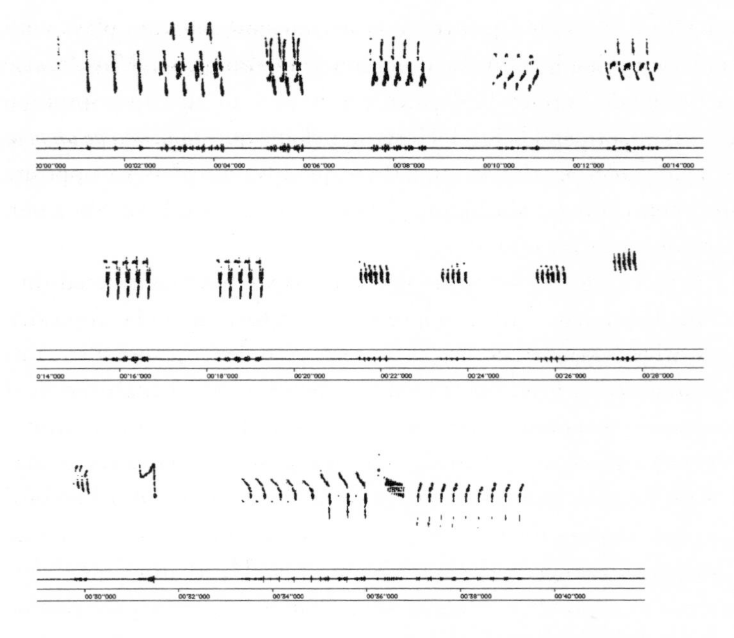

A SONOGRAM OF FORTY SECONDS OUT OF
A SINGLE THIRTY-MINUTE MOCKINGBIRD SONG

some detail one forty-second section of a thirty-minute mockingbird song. This exact sequence of motifs is never repeated again in the other twenty-nine minutes. This excerpt is either one long phrase or two shorter ones divided by a very slight pause. It's a much more compressed sonogram than any we've previously looked at it, because it is trying to cover a much longer stretch of time. This example is a bit less than half a printed page. To print the whole song would take about forty-five pages, with no two phrases exactly the same. But consider, for a start, these forty seconds. Order is clearly visible, like a code waiting to be cracked. To anyone who listens enough to birds to form a catalog of many species' songs in her head, the mockingbird is irresistible, as his song is a compendium of all possible songs. More than a relentless outburst of imitation, it is a charted journey from one surrounding song through the next.

I will try to make sense of the song using all the tools we have considered thus far. First, the pull of language via Garstang and Schwitters. Between each group of syllables is almost the same amount of space, suggesting an overall rhythm, making the whole thing that much closer to music. Each motif recalls the songs of various other birds, but often the mockingbird seems to be teasing one into another. Here we go:

chiķ chu chu chu	4	[imitations of call notes]
ipuchi ipuchi ipuchi ipu	4	[Carolina wren]
chiu chiu chiu	3	[ovenbird?]
pwe pwe pwe pwe pwe	5	[cardinal]
uwe uwe uwe uwe	4	[inching toward blue jay?]
chuwee chuwee chuwee chuwee	4	
piu piu piu piu piu	5	[titmouse?]
piu piu piu piu piu	5	[exactly repeated]

A little more space between, then faster:

chh chh chh chh chh chh	6	
ch ch ch ch ch	5	a little quieter
chh chh chh chh chh chh	6	
si si si si si	5	a little higher
sissiuu	3	[alder flycatcher or warbler coda?]
chii-yuuu	2	[ovenbird slowed down?]
pyah pyah pyah pyah pyah	5	[blue jay]
pyup pyup pyup chi rrrrrr	3–2	[shortened jay into hawk?]
wetu wetu wetu wetu wetu wetu wetu wetu wetu	9	[flicker?]

All right, the syllables are strangely unsatisfying. The sonogram shows regular groupings, each cluster linked to the next like a cipher of sound. What is the connection between a motif and the one that follows? This whole phrase begins with two motifs linked together, 4 and 4, eight notes in total. It ends with a grand statement of (5 + 3 + 2), 10 + 9. I hear and see structure. I don't want to diagram it, I want to inhabit it.

One moment that intrigues me is the shift from the Carolina wren notes to the ovenbird. *Teakettle Teakettle Teakettle,* the wren is supposed to say. *Teacher Teacher Teacher,* the ovenbird. From 3 repeats of 3 to 3 repeats of 2, from *ipuchi ipuchi ipuchi* to *chiu chiu chiu*. The *ipu* in between suggests the fast move—What do these birds have to do with one another? Both are somewhat inconspicuous, hiding in the bushes low to the ground, with impressively loud songs for birds so small. One sings three beats, the other two. How are the songs related? Only by chance in books of bird songs—adaptively, there is little close connection.

I'd like to believe the mockingbird is on to something here. He may be aware of this relationship between how these songs sound. We would do well to learn from his sense of connection, to use his methods to unite the forms underlying the plethora of birdsound patterns. As a composer, the mockingbird has a particular manner in dealing with all he has heard. His distinctive song groups syllables into 3, 4, 5, 6, and at the end of this excerpt, 9—a sequence that will never be exactly repeated. Six minutes later, something similar in shape to the end of this phrase comes again with slight variation: 6 6 5 5 instead of 6 5 6 5, both with the last motif a bit higher. Then a long flourish mixing blue jays and hawks, about the same length, with the same sense of decisive punch but not the exact same notes. The mockingbird is not a machine spewing back rote memorized "packages." He may have committed to memory between one hundred and two hundred syllables, but he puts them together a bit differently every time he sings.

After acknowledging the similarity between the wren and ovenbird songs, this performance continues by working through the two-part *tea-chhr* motif, combined into bouts of four and five then rapidly six, with the syllables in each part of the motif seeming to lean toward the jeer of the blue jay. The mockingbird considers the legitimacy of mixing this tone in, as well as enjoying some balance between repetition, similarity, and contrast. Is this rhythm so exact? I made various transcriptions of this song at different speeds. The more you slow it down, the more new levels of rhythm appear, some which seem far from the shape of the song as it streams by in real time. The scale of this song leads me to trust my ears more than in songs that go by at lightning speed. Unlike Szöke, I prefer a simplified transcription where the musical phrases come clear. It's transposed two and a half octaves down and shows elements that become clear when the song is slowed down to around one-third of its normal speed:

THE SAME FORTY-SECOND PORTION
OF A MOCKINGBIRD SONG TRANSCRIBED
IN MUSICAL NOTATION

With this transcription the time signature also moves from 4/4 to 7/4, passing through 5/4 and 6/4 along the way, then receding to 5/4—a clear movement through increments in space between motifs in addition to number of repetitions in each motif.

Each different human way of noting down the bird's song reveals different senses of patterning within it. What remains most remarkable about the mockingbird's song is the way the composition almost seems to analyze itself. As it proceeds from *teakettle* and *teacher* to *peoow* and *ke chhiiiiir,* the mockingbird's song consists of a reflection of all the bird sounds around and how they resemble one another.

Looking for assistance, I wonder, what would Wallace Craig do? He wrote two hundred pages on a bird song composed of just three notes, proclaiming it the paragon of bird music. He did, however, note that mockingbirds were also pretty fine, though he might have shuddered at a song that takes forty pages just to print out. Would he also hear perfection in the mockingbird's revels? I think he would notice vitality and spirit, with a tremendous will to go on. Perhaps he would hear a tendency to twist one song into another, some evolved need to entwine the sonic fabric of reality. He might say this tendency toward transformation from one imitation to the next, above a ground of rhythms in four, five, and six, would be something "older than the result." The song of the mockingbird thus may have evolved from some simpler, rhythmic bird song like the chipping sparrow's, mixed with the less focused inventiveness of the catbird. Craig would hear it as selection gone wild with prowess, but he would look for a musical reason for the final form, an evolutionary basis for the same beauty that I want to embrace but not quite explain.

Remember, there is no clear reason for mimicry in the avian world. No copied species are fooled. True, it is one easy way to make a song more complex, but the mockingbird goes far beyond that and uses his own special mockingbird rules to compose songs that will not be mistaken for those of any bird but himself, a bird who is better able to fit different bird songs together than any other species alive. He mocks only those who refuse to hear all the possibilities and the sliding interconnections—perhaps that means any species but his own.

Science has been somewhat scared of tackling this species' song, in part because it is so hard to generalize about something so complex. "Many dissertations have been ruined on the mockingbird," warns Don Kroodsma. Just when you conclude that a quality is endemic to the song, the next bird is liable to do something completely different. All birds remain individuals, and mockingbirds are especially singular as they let loose their songs. They live to create.

Some scientists once made the astonishing claim that mockingbirds might be imitating the calls and songs of birds now extinct. If 85% of the mockingbird's song is not composed of easily identifiable imitations, then, they might be the sounds of birds no longer with us; this would make his powerful song not a fanfare but an elegy. Today few ornithologists would support such a view because most mockingbird notes can be traced to specific sounds that the birds hear in their immediate environment. Some claim they imitate clocks and car alarms, but I suspect those alarms have been designed around the same principles that govern the mockingbird's sound.

Who is the song for? Based on its sonic characteristics—loud, clear, aggressive—many assumed that this strong song had a primarily territorial function. But a 1987 study by Randall Breitwisch and George Whitesides showed that male mockingbirds seemed to direct their song into their territory, rather than across their borders to the realms of other males. This suggests that the song is sung toward mates, not rivals. When mated males chased intruders from their homelands, they were assertive but usually silent, unlike song sparrows. Not really what investigators expected.

And yet, one of the finest and most elaborate studies of mockingbirds, a thesis by Peter Merritt, formerly a biologist and now a regional planner in Florida, reveals that the birds singing the loudest and most extensive songs were not necessarily those with the best territory or the greatest mating success. Males achieving the highest reproductive rank were among those who spent the least time singing!

They spent more time mating than advertising themselves. So what's the point of the superior song? Merritt suggests that the mockingbird's prodigious ability to sing may have evolved as a way for males to publicize their learning abilities to potential mates.

Unmated males did sing a bit more than their hitched counterparts. Unlike most birds, though, male mockingbirds do tend to sing just before and during sex itself. Merritt saw the birds do it just six times.

> Prior to each copulation, the female was on the ground as the male sang from an elevated perch. On one occasion the female quivered her wings and the male landed several meters in front of her. The male then walked to within half a meter of the female and then walked around her one and a half times before mounting her and remaining there for several seconds.

He sang continuously until he left her, the song lasting far longer than the actual mating. For mockingbirds, song leads up to sex, and far more time is spent on singing than mating. Shall we respect the song as an end in itself?

Just how do mockingbirds produce such a wide variety of bird sounds? Because of its incredible propensity to copy and re-form, Sue Anne Zollinger and Roderick Suthers at Indiana University realized the mockingbird was an ideal subject for an experiment on the precise workings of the syrinx. While he didn't try to play it himself, Suthers had earlier figured out how to insert tiny wires into muscles surrounding the syrinx to monitor exactly how they function during singing. Delicate surgery inserts the wires and a tiny recording backpack, and when the experiment is done, everything is removed and the birds can return to the aviary. Every song they sing generates reams of data for the scientists to analyze.

Zollinger tutored the baby birds with diverse bird songs known to be difficult to produce. The brown-headed cowbird has already been mentioned as the North American songbird with the greatest frequency range. Its song consists of abrupt gutteral leaps from 200 to 11000 Hz. Cowbirds produce such astonishing jumps by alternating between one side of the syrinx and the other. A mockingbird's range is only from 740 to 6700 Hz. How does he strive to copy a song that is physically beyond his ability? He makes the same switch from one side to the other, but either leaves out the notes he cannot reach or replaces them with those at the edge of his range, a bit like a sitar soloist struggling to keep up with a greater master. Zollinger thinks this shows that the mockingbird can *hear* a wider range than he can sing, and he tries his best to compensate. The closer he emulates the method of the tutor species, the closer he gets to the original song.

Suthers and Zollinger studied the movement of mockingbird muscles in the process of imitating the downsliding *pwe pwe pwe pwe* whistle of the cardinal, produced with a seamless shift from the left side of the syrinx to the right. To our ears the descending sweeping whistle of the cardinal sounds smooth and easy, but the way the bird produces such a pure timbre is surpringly complex. Here's how Suthers describes the process step by step, discovered in an earlier experiment on cardinals: First, closure of the left syrinx; next, expiratory muscle contraction; third, open the right syrinx and set it up to make the first half of the sweep; fourth, close the right syrinx, while fifth, opening the left one and setting it up to continue the sweep; sixth, close the left syrinx to complete the sound; seventh, relax the breathing-out muscles; eighth, contract the breathing-in muscles; ninth, open both syrinxes for a quick minibreath to replace the air used in producing the motif. Then start again—up to sixteen times a second!

The closer the mockingbirds got to copying the cardinal's mechanism of sound production, the nearer they got to its song. Zollinger and Suthers are convinced that this means that the rules of song making are shared among varying bird species. All birds shift back and forth from one side of the syrinx to the other. Syrinxes in songbirds are amazingly flexible, and this is why some birds are able, in new circumstances, to produce sounds far different from what they usually sing.

Mockingbirds are experimenters by nature, so they are incessantly driven to rehearse the relationships between one mode of song production and the next. This bird knows all the routines a syrinx might go through; indeed, his moves might be heard as the Czerny etudes to beat if you want to put your syrinx through a powerful workout. If he can't imitate the whole of something, he takes the parts he needs and recombines them with the patterns he already knows.

What happens when he imitates clocks, car alarms, and barking dogs? He cannot use the same method to make those sounds as their sources do, but he does translate those sounds into new challenges for the syrinx. Why, why, why does he go on in threes, fours, and sixes? Ask him that. Ask him that. Ask him that. That that that. Ask ask ask.

What is the best reason for the presence of form and balance in nature? Although hardly anyone has considered the audible order of natural sound as evidence of this symmetry, there are plenty of accounts of parallels and processes in the visual world of natural structure, from D'Arcy Thompson's legendary *On Growth and Form* in the early twentieth century to Tyler Volk's recent *Metapatterns.* Beginnings, middles, ends. Repeats, breaks, rises and falls. Twos, threes, fours, sixes. What we perceive in nature is not a continuous surge and flow, but discrete things, rhythms, patterns seen and heard. This is how it is throughout nature, as D'Arcy Thompson wrote without really having sound in mind—"The lines of the spectrum, the six families of crystals, the chemical elements, all illustrate this principle of discontinuity. . . . Nature proceeds from one type to another among organic as well as inorganic forms."

These types vary according to their own rules, as constrained by the limits of mathematical possibilities. The laws of bird song too are written in an ancient natural language of number, and the mockingbird reveals these connections as it stretches the Carolina wren's *teakettle* into the *teacher* of the ovenbird and on to the cry of the jay. No wonder the Hopi heard this bird give all other creatures their names and their sounds—his litany of interconnection is as plain as day. Evolution was bound to let at least one species run the gamut of possibilities through its own internal logic of beautiful survival.

The great ornithologist Alexander Skutch, who died last year in Costa Rica at the age of ninety-nine, believed that natural beauty is no mere side effect of evolution. Natural selection, he points out, is essentially destructive. It eliminates those unable to make the grade while offering no special care to those that live on. The constructive part of life is where he sees the glimmer of a final cause: the will of each creature to survive. This need to endure, to live as one must, to sing to fulfill what your genes have given you, is the way evolution feels from inside. This is why the lyrebird cannot stop in the middle of his courtship dance, and why, if you deprive a satin bowerbird of the blue petals he must collect to decorate its mating bower, he will kill the first blue bird it finds and lay its feathers in front of his lair. He simply *must* have blue.

For Skutch, evolution is part of a larger end, a march toward a harmony where all things are connected, a strong living force to contradict inanimate matter's tendency to decay into chaos. Things may fall apart, but every living thing holds its place in the stream. The more science reveals nature's wonders, the more we train ourselves to resist the old intuitions that used to guide human understanding of nature. Plato thought birds sang because they were happy. Today we are supposed to fit all the myriad variations of life into the simple mechanism of evolution. But what guides that mechanism along? The will to live. The emotions of necessity—all those birds born to sing.

The mockingbird's elusive complexity still sounds like a code waiting to be cracked. It may be harder to answer the question of *what* birds sing than why! Think of the power of mantras, Hindu incantations that on the surface are composed of nonsense syllables but when internalized lead the soul to higher spiritual states. So are the sounds still nonsense? The mantra's power lies in the telling and repeating, not in any specific message conveyed by the arrangement of sounds.

Berkeley philosopher Frits Staal realized that this kind of communication has something in common with the way we have come to understand bird song. It is not quite music, not quite language, but rather a structured string of sounds with a clear ritual purpose. Mantras are meant to be repeated by the chanter over and over, approaching endless repetition as the sound swirls inside the brain. You hypnotize yourself with the endless possibility in each abstract sound and the crystallization of the pattern.

Sanskrit is among the oldest of languages, of all our Indo-European tongues. Now Staal says mantras, rhythms of sound that do not quite make sense, may lie at the roots of Sanskrit. Here's an ancient song from the Vedas to be sung in the forest: *Ayamaya mayamayamayamayamauhova.* Literally all it means is "thisonethisone thisonethisonethisonnnnnne . . ." You are supposed to sing it when you consecrate an altar out of doors. Staal believes such resonating, repeating measures of sound may be older than human language itself. It may have worked like this: Our ancestors chanted rhythmic patterns of sound long before we ever thought that sounds should signify specific things. Sound came before sense, before we had history, back in the time of birds. Language came out of ritual rather than the other way around.

Why does nature extend itself for millennia with patterns of excess, beauty, ritual, game? For Alexander Skutch, who spent nearly a century seeking a reason for nature's beauty, humanity's holy grail is to admit that nature is beautiful. As parts of nature ourselves, "we have an inalienable right to judge nature by our highest ethical standards. . . . We find that a large part of its beauty has been promoted by mutually beneficial relations between organisms, of the same or different species, that are morally admirable." Birds settle disputes and cry for love with song, quelling any need for extra violence. The beauty of their sounds is part of a natural harmony that pervades all living things. The beautiful is good. The mockingbird does not taunt; rather, he admits influence, with a kind of respect.

When Partha Mitra, author of the fabulous sound comparison software that Tchernichovski used to decipher all those baby zebra finch sounds, asked me as a musician to help him ferret out exactly what it is that makes mockingbird song seem musical, he cautioned me away from studying these very diagrams his programs produced. "Careful," he warned. "Don't start analyzing, that's what *we* are doing. Just listen, that's how you can tell if it's music." Mitra, a native of Calcutta, owns an intense gaze and several patents related to cell phone technology. His chief focus now is the rules of communication that govern the brain. He's interested in these enigmatic bird adventures because they might teach us how brains put patterns together, perhaps leading to an initial notion of what that higher level abstraction called *mind* might be.

He and his team are at work on the larger scale structures of the mockingbird song. Earlier in this chapter I looked at less than one minute of the song. Partha's after the bigger picture. Over the longer course of a song he's found a general fluctuation of pitch, a bit up, then down. "We have now analyzed four other mockingbirds with relation to the slow fluctuations of pitch that I have described to you. It seems safe to say that the mockingbirds are keeping track of the pitch of their notes over at least a minute's time scale, a very slow, hard-to-perceive melody of sorts. . . . The two Florida birds show a fairly regular pitch-periodicity, while the two Massachusetts birds show slow fluctuations without necessarily being completely periodic."

"Do you think a detailed notation of this mockingbird music will help us make sense of it?" I suggested.

"Listen. Indian music has gotten by without notation for thousands of years, and it is deeply complex. I think the mockingbird is much closer to improvisation than written composition."

All right, I will simply listen. But as a player I do want to join in. How can I tell if I imagine order or if I am finding it in what comes right from the bird? I am not so willing as Tchernichovski to trust a machine's detection of structure derived by comparing the data to itself. Like Dowsett-Lemaire, I still trust listening first.

I want to know how he does it as much as I wish to decode how he goes from there to here. So I've sampled all those mockingbird syllables into a computer and programmed a keyboard so I can play them all together, creating my own patterns out of the material he gives me. Can I learn to play mockingbird music with human tools? The wider our openness to musical opportunity expands, the more music we may hear in the natural world. Playing with the bird's sound material in a testing, improvisatory way, I may get to feel his aesthetic sense before I can define it. It's not as bold an approach as Ayanna Alexander's attempt to master the syrinx, but it may be easier.

By playing with the mockingbird's material I may find a practical way into his mind. I don't expect to convince him that I'm hip to his game, but I want to get closer to his sense of joy, producing Whitman's "thousand random songs an hour, all set to awaken my own song." *Paumanok! Paumanok! Paumanok!* A poem needing words so soon is all too human, but music may range across species lines. One day I hope to hear the mockingbird that lives around my house imitating my imitation of him. Naïvely, I imagine that the more mocker-like my music, the more I might impress my flying neighbor and pique his interest in yet another alien tricky sound, more raw material to work into his grand, ancient composition refined over eons of bird years. That's him all right, you can always tell that's no catbird or thrasher, down by the edge of the field, singing where the weeping willows wave.

CHAPTER 9

The Opposite of Time

ON MAY 27, 1784, Wolfgang Amadeus Mozart purchased a pet starling. He wrote down his purchase in his diary of expenses, along with a transcription of what the bird sang. He heard the starling whistling a fragment of his own piano concerto in G Major, K. 453. Thing is, he had just completed that piece on April 12 of the same year—it had not yet been published! Only a few people had heard the work by the end of May, perhaps only the pupil for whom Mozart had written the piece. How could that bird have learned Mozart's melody? Was it only coincidence? Mozart often frequented this particular shop, and he was known to often whistle in public. Perhaps the bird heard him on a previous visit? Or maybe Mozart's pupil had gone into the store before him.

Even though Mozart's life and notebooks have been pored over by scholars for centuries, the starling story did not much impress musicologists until the two starling scientists, Meredith King and Andrew West, unearthed it about fifteen years ago. Just as the birds they raised at home quickly developed their own starlingesque renditions of "Way Down Upon the Swa" and the fluorescent light hum above their cages, according to Mozart's notebook, his starling had already modified the great man's melody: G natural was changed to a G sharp, immediately making the tune sound centuries ahead of its time.

Nothing more appears in Mozart's papers regarding the starling until June 4, 1787. On that day the beloved bird died, and a full funeral was held, with all the guests wearing solemn attire. The maestro even recited a poem of his own for the occasion:

> A little fool lies here
> Whom I held dear—
> A starling in the prime
> Of his brief time
> Whose doom it was to drain
> Death's bitter pain. . . .
> He was not naughty, quite,
> But gay and bright,
> And under all his brag
> A foolish wag. . . .

Historians tend to view the whole thing as yet more proof of Mozart's immaturity—he held this ceremony for a common, squawky bird in the same week that his father died. But did he ever learn musically from the impetuous irregularity of his bird's singing style?

Later that same June, Mozart entered a score in his catalog numbered K. 522, titled "A Musical Joke," a composition for an unusual chamber orchestra of two French horns, two violins, viola, and bass. The liner notes to a recording of the piece says that "in the first movement we hear the awkward, unproportioned illogical piecing together of uninspired material." Hmmmph. "Later the *andante cantabile* contains a grotesque cadenza which goes on far too long and pretentiously ends with a comical deep pizzicato note." After hearing a starling who learned one of his own melodies, Mozart wrote a piece where he learned from the disjointed, rather unclassical, and certainly nonhuman musical sense that marks the starling's song:

In all the pages devoted to Mozart scholarship there is no explanation of this curious cadenza. It is possible that King and West have solved a minor Mozart caper—just listen to those violin leaps and slides. Repeating patterns, faster and faster, speeding speeding, rolling higher and higher up with a *breezis bleesix bleep!* Making the violin squeal to its limits may be the highest tribute this genius could offer his beloved lost bird. He turned his music away from the familiar energy and exuberance into a weirdness he may have gleaned from the starling's special way with sound.

Human musicians have learned from birds for as long as there has been music. Whether our music sounds much like the bird originals depends on what bird and what music. What is it exactly that bird songs offer composers? The acute outdoor listeners among us cannot fail to take the forest songs of dawn and dusk as serious opportunities for inspiration—their freedom from set musical constraints, their suddenness blended with predictability. Because it fits so clearly into nature, bird music seems always somehow *right.* One cannot deny its importance and necessity to the flow of life, whereas every human construction seems to be always fighting to assert its own importance. Artists too have their territories to stake out.

More melodious birds are much easier to accommodate. Antonio Vivaldi wrote his famous flute concerto "Il Gardellino," the Goldfinch, in 1702. It is one of his earliest pieces, op. 10, no. 3. The following excerpt from the solo flute part shows a clear similarity to Kircher's nightingale transcription, though certainly tamed to comply with turn of the eighteenth-century classical aesthetics. It's all in a joyous major key. The birdlike trills are rigorously in time:

Vivaldi got a theme and structure from the finch and placed it inside the musical conventions of his time. To accurately record how and what birds sing, musical notation has to be stretched to its limits, as Mathews and Thorpe have shown. When we write bird music down, we always abstract from it. We learn the limitations of our tools in the emulation of nature.

Beethoven's famous Sixth Symphony, the Pastoral, is one of the most well-known classical pieces said to derive from the sounds of nature. The composer originally planned to provide detailed program notes to help explain the symphony, but later decided that "anyone who has an idea of country life can make out the intentions of the author, without a lot of titles." Nevertheless he does name the first movement "Awakening of Joyful Feelings upon Arriving in the Country," a rather specific title. The music is airy, tonal, light like a whoosh of fresh air. The opening theme is cheerful and birdlike, although not coming from any bird in particular. There is little modulation from one tonal center to another, giving the impression of the resonant incessance of landscape, resounding on and on without beginning or end.

At the end of the second movement, "Scene at the Brook," there is a specific trio passage where flute, oboe, and clarinet play melodies derived from the songs of the yellowhammer, quail, and cuckoo. This is the most well-known evocation of bird song in nineteenth-century music, and one of the first to attempt a kind of ecological interaction among different species singing at once. Rather than admit any sense of respect for any music hidden in wild birds' song, Beethoven later claimed the passage was intended as a musical joke. His vision may have been too complete and self-contained to permit him to honestly accept what real nature might offer him.

What makes a piece of human music birdlike? Short disconnected phrases perhaps? Reaching, yearning trills, blends of noise and pure sounds? Brief melodies that carry far through the trees? A good ornament or a fine armament? We applaud how music has become sufficiently open in this century to encompass recitations and flourishes that clearly show an effort to learn from the genius of birds. This is an interesting sign that music may have progressed—not that it is "better" now than in Vivaldi's or Mozart's time, but it is more free, able to open up to the cycles and moods of the natural world and work these irregularities into what we sing and play without needing to bend what's out there into constraining human rules. What once sounded like a joke now becomes a direct source of inspiration.

Most of the major composers of the Romantic era wrote pieces based around themes that they said derived from bird songs, but their music is more likely to place the exuberance of birds inside the musical language of the time than to bend the rules of the music by listening ever more closely to nature. There is one composer in the classical tradition who heard more than charm in the music of birds. This is the French composer Olivier Messiaen, one of the most distinctive voices in twentieth-century music.

Like the strange blending of nightingale and bomber recorded by the BBC, this collaboration between human and birds began in the midst of war. The twenty-nine-year-old Messiaen was on dawn watch in the French army in 1940, stationed in Verdun. The sun was rising and all the birds began to sing together. "Listen to them," he told fellow sentry Etienne Pasquier, a cellist, "they're giving each other assignments. They'll reunite tonight, at which time they'll recount what they saw during the day." Messiaen had been transcribing bird songs since he was a teenager, but he had not previously thought to use them seriously in his music.

There was also a clarinetist in the regiment, an Algerian named Henri Akoka. After many days of dawn watches Messiaen began writing a solo piece for him, "Abyss of the Birds." Before Akoka had time to try it out, Germany invaded and the musicians were captured in the forest and transferred to Stalag 8A in Germany. It is here that Akoka first had a chance to sight-read the piece, with Pasquier serving as the music stand. "I'll never be able to play it," Akoka complained. "It's impossible." Some of Messiaen's birdlike notes were incredibly high in pitch, with precise and irregular rhythms. "Yes, yes, you will, you're getting there, you'll see," Messiaen assured him. All three musicians were prisoners, but they had already become part of the genesis of one of the finest pieces of chamber music ever written, Messiaen's "Quartet for the End of Time."

A German officer in the camp, Karl-Albert Brüll, heard of Messiaen's prowess and made sure he was provided with music paper. A violinist, Jean Le Boulaire, was soon interned at the camp. The Red Cross provided a few musical instruments, although there were 30,000 prisoners and only a few violins and cellos and just one piano. Akoka had managed to retain his clarinet. Messiaen persevered, and while a prisoner he completed his most celebrated composition, the first of many to use the sounds of birds. In eight movements of astonishingly beautiful music, the four instruments weave in and out though unusual harmonies, rhythms derived from ancient Indian and Greek music, and impressionistic chords that rise up toward a sonic Heaven. In the first movement the clarinet and violin trade sounds from blackbird and nightingale, and the solo clarinet third movement is a musical attempt to link the endless enthusiasm of singing birds with the long, dark weight of eternity.

"The birds," Messiaen wrote in his notes to the piece, "are the opposite of time. They represent our longing for light, for stars, for rainbows, and for jubilant song." As a devout believer, he did not think of the end of time as a holocaust but as the dawn of an Eternity where Jesus will console us. Yet in the resplendent harmonies that draw the listener upward—the violin and piano finally rising higher and higher into the farthest reaches of human hearing—you can hear any Heaven you wish for.

The piece was premiered in the stalag on January 15, 1941, in the midst of the most terrible war humanity has known. Hundreds of prisoners and their captors were in attendance. In the middle of this battle that tore the world asunder, it was still possible for men to make music that will endure, perhaps longer than any of the terrors of the time that are still so close to us. We remember Beethoven and Mozart much more clearly than the Napoleonic wars. Only the sound of birds lasts longer than the greatest music.

Messiaen is right in saying that bird songs are the opposite of time. Their patterns offer constancy through the generations of human composers who have tried to make use of them. A truly innovative and experimental composer, he knew something about nature beyond its role as a source of new sounds. He knew how to listen:

> In my hours of gloom, when I am suddenly aware of my own futility, when every musical idiom—classical, oriental, ancient, modern and ultramodern—appears to me as no more than admirable painstaking experimentation without any ultimate justification, what is left for me but to seek out the true, lost face of music somewhere off in the forest, in the fields, in the mountains or on the seashore, among the birds.

Unlike earlier and later composers, who often tried to emulate the musical energy of birds by using the shapes of their melodies, Messiaen was as concerned with the tone quality and rhythm of birds as he was with the tunes. His massive, seven-volume *Traité de rythme, de couleur et d'ornithologie* includes two six-hundred-page books (vol. 5, parts 1 and 2) of detailed transcriptions of bird songs, most done right in the field with no recording technology. If the bird sang irregularly and haltingly, he would transcribe the song that way and adapt it so that human musicians could make use of this irregularity. And by blending together all the bird songs that might be heard in one location, he combined exact notations with ecological ideas.

In the sprawling solo piano work "Catalogue d'Oiseaux" he uses the single instrument of the piano to play an entire treatise on the sounds of birds and their habitats. Each bird is conjured with precise harmonies, rhythms, and situations, with the phrases of birds placed together based on time of day, shared environment, and surrounding sounds—those of water, dawn, night, frogs, wind. The music is abstract enough to be far from the easy programmatic sense of Beethoven, but it is at the same time both more cryptic and more exact. The thirteenth and final movement, "Le courlis cendré," is based on the song of the Eurasian curlew, a large curve-billed sandpiper. Although not generally heard as an especially musical song, it is for Messiaen a paragon of avian song structure, with a series of rising gull-like sighs, then a series of quicker, harsher notes of increasing complexity. Here is his transcription made in the field, complete with strict dynamic markings:

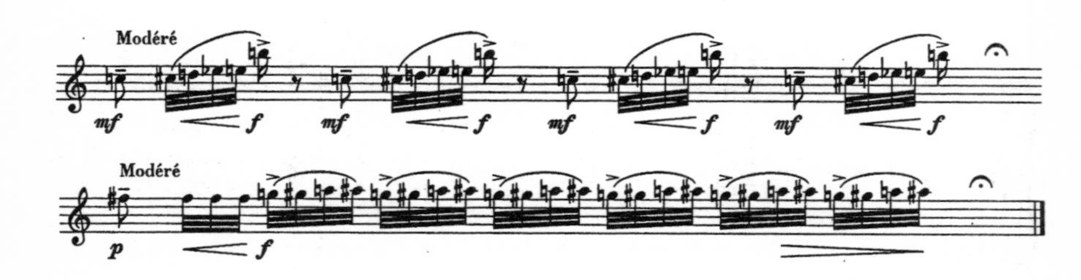

For an alternative visual way to present the same sonic information, here is a sonogram of the same sort of bird:

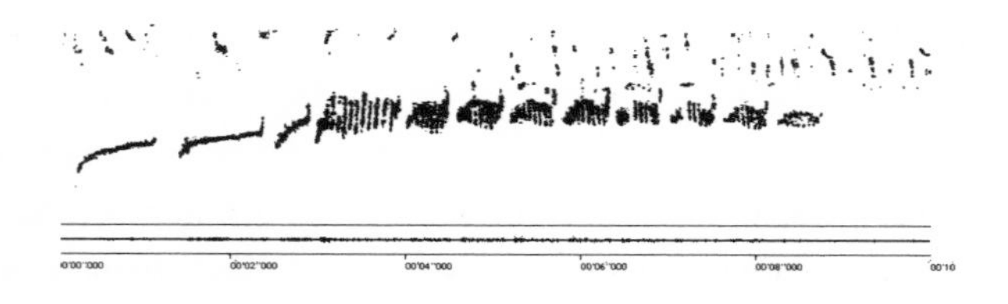

In his composition based on this song, Messiaen slows it all down and emulates the timbre with dissonant but open harmonies. In this conclusion to a huge, magisterial work, the composer uses the piano's dark harmonies to get at the longing essence of this rising call of mystery. Here are two fleshed-out piano excerpts based on the transcription on page 196:

Out of the whole two-hour "Catalogue d'Oiseaux," to me this final movement has the most sublime clarity. Here's why he needs to slow it down:

> A bird, being much smaller than we are, with a heart that beats faster and nervous reactions that are much quicker, sings in extremely swift tempos, absolutely impossible for our instruments. I'm therefore obliged to transcribe the song into a slower tempo. Moreover, this rapidity is combined with an extreme shrillness, for birds are able to sing in extremely high registers that cannot be reproduced on our instruments; so I write one, two, or three octaves lower.

Listening to Messiaen, you hear a mind deeply involved in the world of birds, trying to draw you into their unique music and the immediacy of the natural world at the same time.

"I am an ornithologist," said Messiaen, "and a rhythmician." His ideas about rhythm are unique. "Rhythmic music," he writes, "is music that scorns repetition, squareness, and equal divisions." Instead, "it is inspired by the movements of nature, movements of free and unequal durations." He hears rhythm in the sun rising and setting, the great choruses of birds, the uneven evenness of a rushing stream. When we march in unison or dance in a club to the same techno beat, we are far away from rhythm in Messiaen's sense. For him, the greatest of classical rhythmicians are not those who exaggerate the beat but those who stretch it most subtly, like Mozart, whose rhythm is also far from the monotony of repetition.

Messiaen traveled the world transcribing the most complex bird sounds, many of which made their way into his music. Syd Curtis, the greatest living expert on lyrebird song, heard of Messiaen's attentiveness to bird song in the 1970s and sent him some recordings. "I had tape-recorded the lyrebirds with the famous flute song, and I reckoned that Messiaen would be interested to hear birds that used human music in their songs—turning his own approach on its head, in a way." The maestro responded that he had to hear the source of this incredible song in his natural habitat. He arranged a tour to Australia.

Curtis met him and his wife, Yvonne Loriod, at their hotel in Brisbane just before sunrise and drove them an hour to a place where the Albert's lyrebird could be heard. Curtis found Messiaen's listening ability more impressive than his music:

> What a great ornithologist was lost to the world by his concentrating on music! My lyrebird sang for him at dawn, and so did a number of other local species. I held a flashlight while Messiaen wrote busily on manuscript paper. Madame Messiaen kept him supplied with sharp pencils. My wife and I stood by and watched.

After twenty minutes or so, no new birds sounds were being made, and he decided he'd got them all down. So he turns back to the beginning and goes through his notation, whistling or singing with such accuracy that I had no trouble in identifying the species for him. What a fantastic performance! He had never been to Australia before. They were all bird sounds that he had never heard before. And of course, none of their songs fit our human diatonic scale or strict rhythms.

Messiaen's final piece, premiered in New York just after his death in 1992, is titled "Eclairs sur l'Au-Dela" ("Visions of the Hereafter"). The third movement, "L'Oiseau-lyre et la Ville-fiancée," is based on what he heard in Australia, although it is a superb lyrebird, recorded outside of Melbourne later in the trip, not the more elusive Albert's. Here is Messiaen's transcription of the mimicry performed by this bird:

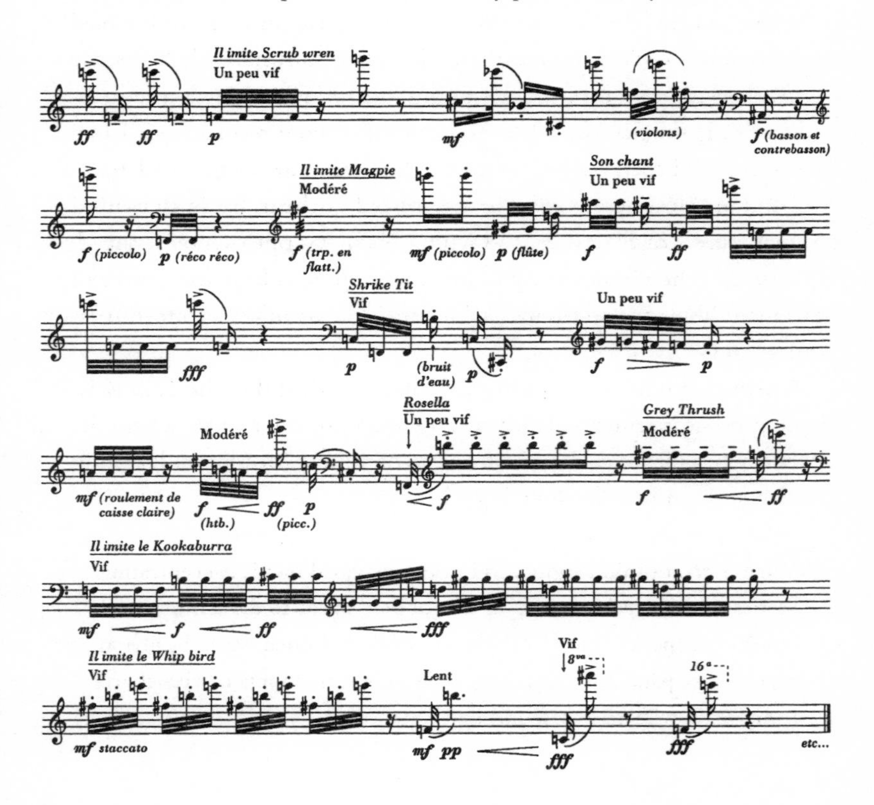

Yvonne Loriod later wrote to Curtis, "Perhaps some day you will hear a recording of this work, and you may be able to recognize your dear friend the lyrebird, who illuminates the whole section, and this thanks to you." An honor, said Curtis to me, but he doubts any lyrebird would recognize his own music in what Messiaen has done with it!

The rich musicality of bird song always held more for Messiaen than what he could derive from it. When he was asked if nature does things better than civilization, he replied, "I dare not answer—my response would be that civilization has spoiled us, has taken away our freshness of observation." We live amid so much baggage that it is hard to clear it all away and simply listen to the natural world around us. Why bother? What do we get from it? How will it help us or our culture progress?

The answer may lie in Messiaen himself. He used bird songs in two distinct ways, first in attempts to offer an exact musical portrait of a series of species and their intertwined actions and habitats. Second, he saw the rhythms and structure of the songs as musical material, something to bend and twist in creative ways, transforming the scores on paper the way later composers manipulate sounds electronically.

In the last century we have at least been opened up to the musical possibilities of all sounds. I play these bird songs to some of my students and they find the scratchy riffs of the nightingale all the more musical because they produce the rhythmic noises made by DJs and electronica alchemists, where sounds draw the listener in because they have not been heard in this context before. As human musical preferences meander and evolve, we find ever new ways to bring the bird world in, yet each time the birds and their inherent mysteries remain beyond our reach.

Our ears cleansed and prejudices cast aside, bird sounds should sound more musical to us than ever before. Why then does Messiaen lament the fact that so few of his students and followers appreciated his attentiveness to birds? Perhaps he lived a generation too early for the ecological value of his work to sink in. It is still a minority of musicians who take the time to hear the musicality of nature. There is a danger in this listening to the endlessness of the wider world: If bird song is complete in itself, the more we listen to it, the less we will want to make music at all. The more time composers spend with them, the more beautiful they sound, and the less there is for the composer to do. Attending to the music of nature has led more than one composer to question his own calling.

Scottish composer Magnus Robb has written some of the most precise bird-based works since Messiaen. His piece "Sprosser: Hallucinations of Purity," composed in 1998, features a percussion part based closely on the rattles and rhythms of the thrush (Sprosser) nightingale, with rhythms meticulously notated, resembling the transcriptions of Sotavalta. The string parts, for violin, viola, cello, and bass, are closely based on the harmonic reachings of a vastly slowed-down hermit thrush. The result is a confluence of two radically different kinds of bird music. What is the purity the piece is dreaming toward? That necessity our own music will never reach.

When composers transcribe bird songs, they are looking for new sources of stimulation and new ways to discover structure and sense. This is not quite the same reason that scientists transcribe the same songs. As the contemporary French composer and "zoomusicologist" François-Bernard Mâche puts it:

> If I acknowledge that the analysis of bird song is a help in my reflection as a composer, I will soon be suspected of being more imaginative than a scientist should be. Where I think I encounter refrains, anticipations, and reminiscences, the biologist is satisfied with classifying equivalent signals . . . as if musical play were a human privilege, and no similar freedom could be imagined for the animal.

When I tried to contact Magnus Robb about his music, it was more than a year before I received a reply from him in Amsterdam:

> Sorry I have taken so long to respond. . . . But my main work now is as a travelling bird-sound recordist, part of a team working on a new

> sound guide to Western Palearctic birds. So far we have amassed a collection of about 21,500 recordings and my own contribution has been more than 12,500. I am also the one doing most of the editing and studio work, so I have very little time left for composing. I enjoy this work very much, and I do see it as an extention of my previous creativity with sound. Composing was always a bit of a struggle and the rewards usually very meager and brief—once in a while getting to hear a premiere. I derive greater pleasure just now from the unabstracted bird songs and sounds themselves, although I often get sidetracked by ornithologial aspects where aesthetics don't play such a strong part. . . . Although in Kazakhstan this year I recorded the Black Lark which claps its wings while singing a highly complex song, thereby betraying its own analysis of the rhythm . . .

Birds may know their own secrets. This quality latent in all bird sound may be the reason why they are so popular as raw material by composers of *musique concrete*—that kind of music made entirely by transforming sound on tape or in digital form with the use of computers. Although originally a kind of experimental music pioneered in France by Pierre Schaeffer and Pierre Henri in the Group for Musical Research, the method of composing entirely out of prerecorded sound is now an essential tool that can be produced at high quality on notebook computers for today's popular music.

The composer and sound designer Douglas Quin has composed and improvised pieces based entirely on sampled bird sounds for several decades. The more technology advances, the easier it is to perform live concerts with these sounds, rather than laboriously transforming them on earlier tape machines. Quin has set up banks of sound samples on electronic instruments, where each musician triggers bird sounds with their own electronic controllers, some keyboards, others based on drum sets or guitars. His piece "Ice Diver" combines longing, evocative loon calls with a flutist playing transcriptions of portions of those same songs, blending pure, resonating harmonies with the ethereal paean of the loneliest bird in America.

Quin has also used F. S. Mathews's song transcriptions as scores for a clarinetist to perform over sampled backgrounds of natural "ambiences" from different ecosystems around the world. One movement of his "Yasashii Kaze" ("Gentle Wind") begins with that famously weird *whoo ooo ooh* of the screech owl; another begins with the typical "twoness" of the brown thrasher, moving on to the whortle of the robin and the optimistic battle cries of the song sparrow, with much of the piece being a guided improvisation for the clarinetist based on Mathews's melodic fragments. The original material is slowed down enough so that the original transcriptions are much more sources for inspiration than direct quotations. "Remember, Mathews had no access to tape recordings or sonographs to help him make sense of these beautiful bird songs," Quin told me. "With all our technology, and the inundation of media we enjoy today, we have lost that perceptual acuity."

In this way Quin's music does not copy bird songs but builds its own logic upon them. This is just what the birds do. Even the marsh warbler does not just copy every bird it can hear, but works these imitations into its own unique flood of sound. As a result, marsh warbler songs are only of interest to other marsh warblers (and an intrepid few human listeners and wonderers). Starlings grasp only that part of Mozart that fits their musical sense, and the same for Mozart trying to learn from a starling. Mockingbirds do not mock, but put all that they hear into their own style, their own music culture, their own genre. Everyone takes on the world of sound for his own sonic possibilities.

So much for slowing down the time frame of birds; what of speeding up human tones to enter their world? The San Francisco–based electroacoustic composer Pamela Z has written a solo piece entitled "Syrinx" that combines the recorded sound of her own voice with the sounds of several songbirds—Bewick's wren, Cassin's finch, and the black-headed grosbeak. "I time and pitch-expanded the very quick bird sounds, enabling me to hear their surprisingly complex melodic material. Then I learned to sing it." The six-minute work transforms her own vocal sounds and those of the birds into each other through computer-based digital sound design. By the end you can't tell what sounds come from her larynx and which come from the birds' syrinxes. The human moan becomes a bird *plink,* and the bird *sneeps* become human sighs. It is an active, burbling meditation between the human and avian worlds.

You hear the songs successively slowed down and sped up, and it sounds exactly like what the mockingbird does for fun. Pamela Z's swirling sighs rise and fall, rhythms come and go, repetition is extended and stretched out. The slowdown techniques pioneered by Szöke are turned into a musical tool. The borders between species are dissolved in the musical exploration of a vocal theme that sounds like *ooowa oowa hoooooo!* Speed it up enough and raise the pitch, jump back and forth, woman and wren are one. Why must they both sing? By this point it's more Zen koan than question. The more we articulate exactly what is done, the less reason we can find why one kind of song makes more practical sense for the bird than any other. And the same goes for human composers, who have come ever closer to inhabiting the wiles of this outside music, to the point of meticulous deconstruction or ultimate awe, giving up the writing of music in order to blithely listen.

And what of musicians who want to play along? Bird fanciers have delighted in teaching new tunes to young birds during the sensitive period when they are still able to pick up something new. When it comes to playing music with an adult bird whose repertoire is fixed, the interaction becomes a new kind of musical experiment. I hearken back to the duet with the captive white-crested laughing thrush with which I began this journey. I play along with birds to enter a new kind of improvisation, with a musician from another species. I want to engage, to interact, to create together, not to sign my territory or find a mate. What does he want? Who does he think I am?

As I played through the net mesh cage, shocked by each new phrase the bird would swing me, I imagined this was a male singing for the usual dual male reasons to sing. When I read later that this is a species in which the males and females sing precisely structured duets, then the whole thing took on another light. Suddenly I could imagine myself part of a pair-bonding ritual, where my phrases were interpreted by the bird within an exact two-part regimen. Science illuminates the musical experience.

Frederic Vencl and Branko Soucek found in the 1970s that male and female laughing thrushes each have a different musical "program"—that is, each decides which of twenty-five different syllables it will sing next after hearing its mate sing a preceding syllable. Their decision trees and transition matrices show a definite pattern that could be reasonably modeled on the vintage computers they had in those ancient days. In fact, it is remarkable that in the thirty years of computer advance since these studies hardly anyone has conducted an analysis of bird songs into programs as elaborate as this, even though any notebook computer has more than enough processing power to do this kind of work. Vencl and Soucek found a definite order and structure to the duets that they were able to simulate with simple computer programs. They described the male and female birds as each having a distinct song program that determines what they sing. Bird as machine. What do they think all this exactness means?

> At this time, it is speculation to suggest what the birds are saying to one another. We have noted that certain syllables are given more often under some conditions. For example F_2-M_{23} seem to be called up after a disturbance such as a loud noise. This reply loop might mean something like, "I am OK, how about you?" $M_{6/22}$-F_1 may represent a synchronization which "entrains" both birds for further singing. It may signify, "I'm ready to sing, are you?"

The practical value of this structured duetting, the authors surmise, must lie in the fact that there is a plan to it all. The birds need to stay in touch with each other in very specific ways.

What does this have to do with my jamming with one in an aviary in Pittsburgh? Now I listen back to the recording of what we did, and then print out my bird–human duet as a sonogram. Smack in the middle of our song, the bird begins by delivering a characteristic *dee to deeto dee to dee to deeep* and I respond with a rising arpeggio and fall. Then he repeats his same riff, as if daring me to get it. (The clarinet is the lower, gray phrase that goes up and comes down. The bird's phrases are similar alternating patterns to the left and right.)

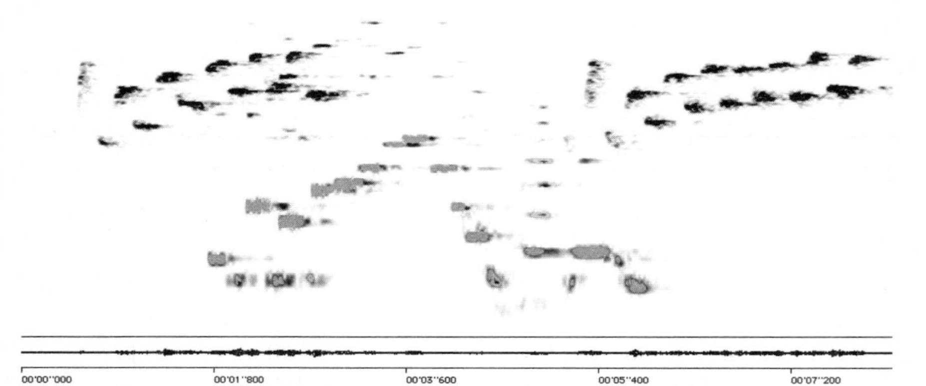

If you're used to reading these diagrams by now, you'll see that both the clarinet and laughing thrush tones make much simpler, clearer marks than many of the noiser songs that have been printed out on these pages. Maybe that's why so much easy music evolves between us. Later I try a descending, slightly syncopated bluesy scale. The bird seems to be trying to match it with high back-and-forth whistles, those parallel marks above mine. Does he know I'm not a bird? Most likely he does. Does he hear my clarinet tone as being related to his own? I suspect something about my instrument's overtone series gets to him. Whether I am rising, falling, or slinking down a scale, the bird finds a way to insert his characteristic phrases along with mine. I don't feel he is trying to jam my signal or intimidate me into silence. Will I ever know if this is more than an anthopomorphic hope that someone else in the natural world likes what I play?

Music is made with a mix of control and daring, answering in sound at the moment more questions are raised. That's the nature of

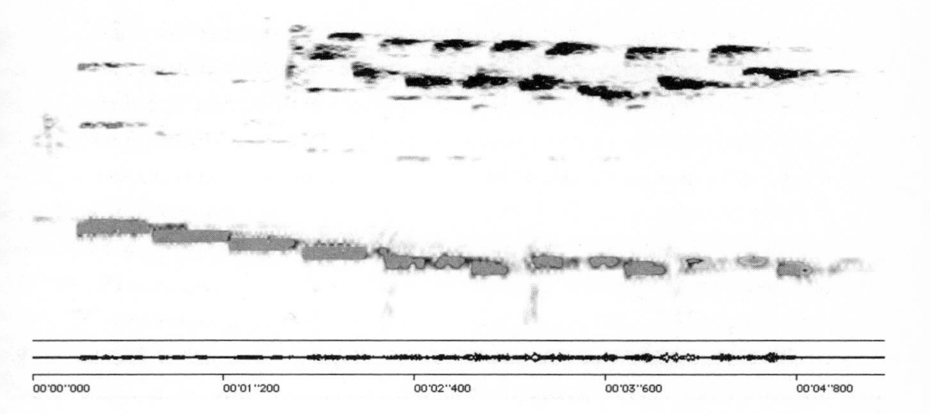

BIRD SINGING IN PARALLEL TO DESCENDING CLARINET SCALE

art inside nature: we will always hear more music than we can find reason for. Even a tiny bird brain can be attuned to the magic of organized sound, where the form might be greater than the function, beauty ever resonant and present, long before we are able to learn from its ways and hear the opposite of time.

Some who have heard my interactions wonder why I don't just try to copy the bird, to show I admire what he sings. I remind them that music is not at its best as a battle or copycat imitation. We learned from song sparrows that it is most aggressive to parrot back to a bird with exactly his same song. Bending what you hear, weaving it with your own songs—now that is suggestive of much more respect. Learn from the mockingbird.

I longed to play more with real live birds, and this time I wanted to do it in the wild. After reading of Messiaen's adventures down under, it seemed that Australia was the place to go. Maybe Syd Curtis could help me too. At first he was by no means enthusiastic. "Let me tell you what happened the one time I played an Albert's lyrebird back a tape of his own song." Here's how this technology-aided instance of interspecies communication went down.

Scene 1 (0:0–2:11) *Alarm in the forest.* A litany of threat and alarm calls, some his own, some from other species, including a small frog and a farmer whistling to his dog. Scene 2 (2:12–2:38) *Threat and terror.* A minute and a half of nothing but the lyrebird mimicking the whistle of a grey goshawk, the most fearsome predator in the forest. Periods of ominous silence. Then, the pitch of the cry inexplicably rises. Scene 3 (3:26–4:52) *Disaster.* A repeat of the multispecies threat calls of scene 1, followed by an increasing intensity of loud, squawking alarm calls, higher and higher, an ultimate of bird anxiety.

"Now that I know what was happening, and the stress I was causing him, I would never play a lyrebird back his own song again. And in 1977, as assistant director of the Queensland Park Service, I drafted a regulation effectively making it illegal to play a tape of a lyrebird song in any Queensland national park." So much for the chance to try the usual playback experiments down under.

"Sure," I told him, "but I won't be copying his song exactly. I'll be offering him my own, as a kind of gift, not a challenge."

"I don't know," he mused, "there's no precedent for that kind of activity. You'll be on your own."

"Will you take Pestel and me to hear an Albert's lyrebird, as you did with Messiaen?" I asked him.

"But of course," he wrote back. "When do you arrive?"

"The first of next month. What if we decide to play along?"

"Well," he thought for a moment. "I'm retired. I won't turn you in."

CHAPTER 10

Becoming a Bird

"MICHAEL, I'VE FOUND OUR BIRD. He's only ten thousand miles away." Pestel had also been looking to play along with a lyrebird in the wild. "Syd Curtis told me that there was only one bird who wouldn't run from us. His name is George, and he lives in the Lamington National Park in Queensland. We have to go soon, while mating season is still on." I rescued Pestel from his warren of an office at Chatham College, where he was cleaning out his studio for the summer. What to do with his huge revolving wooden sound sculptures, or his several rusting innards of pianos, all their dusty strings "prepared" with screws, pencils, and twigs? "Oh, I'll just stuff them into the back of my pickup truck," he decided. "I'll deal with it all when we get back."

We traveled to the other side of the world to keep an appointment with one special bird. Pestel and I needed to find the Albert's lyrebird named George because he is the only member of his wary, elusive species who can stomach the sight and sound of human beings. Any other member of his tribe would dash silently through the underbrush at full speed away from the first sound it heard of a clarinet or flute. George, though, is another story. He has been studied in the wild by two men—an ornithologist and a photographer—for twenty-five years and has learned to tolerate all sorts of strange recording and filming equipment. What would he think of the music of our species? So far no one had asked him.

On a plane ride halfway around the world you lose a day and you hear all sorts of things about whoever's come with you. When he was sixteen, Pestel told his mother he was going to the Expo in Montreal to buy some boots, but he ended up hitchiking across Canada pretending to be a German student named Fred Nietzsche, with a sign that read BERLIN TO FRISCO. When he was picked up on I-5 heading south from Vancouver by the head of the German department of the University of Washington, he knew the jig was up. But he never did go home again.

Twenty-four hours after we took off we were met at the Brisbane airport by the indefatigable Syd Curtis, seventy-six years old, sporting a tam o'shanter and a huge parabolic microphone on a tripod, looking like an overgrown leprechaun. There was a glint in his eyes—he's grateful to find two foreigners eager to share his obsession with one special lyrebird who has been tracked for a quarter of a century. This is the same man who picked up Olivier Messiaen at his hotel room at 5 A.M. to take him to hear an Albert's in 1988, when the great composer and bird song transcriber was eighty. We would retrace musical history.

On the wall of Syd's Brisbane house is a photograph of a male lyrebird with a circle and a slash through it right over the bird's face—the international symbol for *No*. I asked his wife Anne about this. "I simply *hate* lyrebirds," she laughs, "hate them. Had to. Otherwise things would have been too easy for my husband over there."

When Syd, at the age of forty, discovered that male Albert's lyrebirds did their dances atop low-hanging vines instead of the cleared mounds favored by the more numerous and less wary superbs, he ran excitedly to tell his mother, who lived on nearby Tamborine Mountain where Syd grew up. "Ma, I found where the Alberts do their mating dance, no one has ever seen it before." His mother sighed, saying, "Oh, you mean on top of those twigs and branches, yes, I saw that many years ago," and Syd laughed. His mother, a friend of Hartshorne's and a noted ornithologist herself, never thought it interesting enough to tell anyone.

We drive the few hours up long winding roads, past vineyards and villages. The higher we go, the thicker the forest becomes. We enter the national park that was under Syd's jurisdiction for many years, and the road ends at O'Reilly's Rainforest Guesthouse, the famous lodge that has introduced tourists to this landscape for decades. We meet the photographer Glen Threlfo, staff naturalist at the lodge, who gives nightly slide and video presentations of the wildlife and waterfalls found here. Over the many years he's spent in the forest tracking George, Glen has built blinds in the thickets to photograph him in secret, stalking, listening, but also trying to become his friend.

The next morning we awaken an hour before dawn and start walking down the sketchy trail that leads into George's territory. When I first heard of George, I imagined he was one of those somewhat tame birds who live just outside of the famous park lodge. In fact George's territory is a few miles away, down a long and unmarked path. At first he was as wary as his compatriots, but over time he has gotten used to our kind and will not run away at the click of a camera or the whirr of a tape machine. When Glen and Syd first found him, George was already mature, so he is at least thirty years old, the oldest lyrebird who has been followed in the wild, and he shows no signs of slowing down.

Making our way carefully over fallen trees and through tangled underbrush, we walk in near darkness into the piece of the forest held by our bird. At first all is quiet. Then we hear the telltale territorial song high in the canopy. *Breep, booua, bwe, ba boo pu tee! Breaap, booua, bwe, ba boo pu tee!*—a slightly speeding up descending and final rise to a phrase worthy of Claude Debussy in his famous flute piece "Syrinx." We wonder whether, as he sat in the forest meticulously transcribing bird songs, Messiaen ever felt the urge to whistle, sing, or play along on some kind of portable organ or harmonium, adding to the perfect splendor of the songs of birds. I suspect his reverence for nature kept the idea far removed. We have no such pretensions. I quietly click the latch and take the pieces of my rainforest wood clarinet out of its case.

Then, just as Syd predicted, after about fifteen minutes George descends from the trees. He finds one of his branch platforms to begin his Albert cycle of mimicked songs. The sounds commence, in the order and form used by all the Albert's lyrebirds in this district, precisely cultured and different from what other groups do. I cannot believe my luck. He's picked a site just a few meters from my tape recorder. I feel like Ludwig Koch getting a rare song recorded for the very first time. George arches his lyre-shaped glimmering tail feathers right over his head into a curved dome of gossamer maroon. A single red back feather sticks straight up from his behind, a sudden bright surprise. The display begins, with a soundtrack composed of the sounds of all the other birds who inhabit this forest. It is winter, so most of them are silent. At the moment, we have only George to introduce their songs to us.

First, a *sneep* of the crimson rosella parrot, then the *plink chee chee chee chee* of the tiny yellow robin, then a high *pink* of a green catbird, then the harsh *braaaa* of the paradise riflebird. Next, a phrase of descending whistles and scratches from the satin bowerbird, that famous species who so loves blue. *Ba bo de poo traaaaaawh tete pu traaaawhh aaaarrrhh*. In our language it looks like nonsense but in George's voice it is strictly controlled: *Chik. Arrh.* The Aussie king parrot, then the crimson rosella again, and back to more lilts of the satin bowerbird. *Ah errrh ah eh a eerrrrrrrhhh.* The sound of a beak tapping on a branch, but reproduced with his voice, not his beak. *Broo ah ha ha,* the laughing kookaburra. *Cree* of the Lewin's honeyeater. *Plink chee chee chee chee* of the yellow robin again. *Aaaarrrhh* of the bowerbird. Parrot *chiks*. Flapping wings emulated as song, then faint burbles of a white-browed scrub wren. *Mraaah* of the riflebird. A break and a turn for the territorial call: *Breaap, booua, bwe, ba boo pu tee!* Back again to the rosella *sneep*. The cycle continues, with only slight variation.

It's an organized run, a clear composition, not of easy musical tones but of harsh *squeeps* and *braophs*. Lyrebirds take about five years to

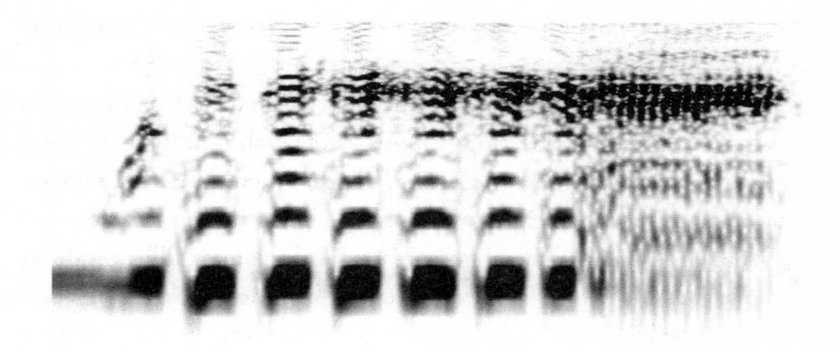

GEORGE'S HALF-SECOND RENDITION OF
ONE NOISY BOWERBIRD NOTE

mature and to refine their songs, making them the bird with the longest sensitive period when song can be learned. L. H. Smith, author of three books on the lyrebird, recalls hearing a superb lyrebird refine his imitations, working them over and over until he was satisfied with his own version. "I used to wonder why he persevered. Unless he was somehow conscious of the difference between good notes and bad ones, why should he have continued practicing?" Through trial and error the bird approached his perfect rendition of the model bird's sound, be it parrot or whipbird or kookaburra. How did he know when he got it right? Lyrebirds choose a specific fragment of another's tune and shape it into their own. They have a fine memory and a distinct aesthetic sense to know exactly what part of another bird's song they want to use.

The lyrebird's song is a clearly composed piece of alien music made out of all kinds of crazy noises. The squawks derived from the calls of the satin bowerbird are among the most complex sounds we have heard from any bird. Above is a sonogram of one syllable, half a second of rich overtones segueing into noise: *eerrrrrrrhhh*. "This may not be easy," says Pestel as he takes in this command performance. "What does George need us for? His music is complete in itself." Michael takes out his silver flute, along with various hardwood bird calls that make any number of species-specific noises. They are carefully attached to his flute with a collection of rubber bands, so all are close to his mouth when he plays, making it easy for him to quickly jump from one call to the next, just like George. I put the pieces of my much simpler clarinet together. Instruments poised in hand, we wait breathlessly for just the right moment to join in.

I think back on the long trajectory that got me here, all these years after I began this project with the surprise of hearing a musical reaction from a white-crested laughing thrush in the Pittsburgh aviary. Back then I was astonished and intrigued. I hoped to make a music that crossed species lines. Only the soul and form mattered.

Over the last five years I have read far too much. I have delved into the history and the poetry and learned a whole new language of scientific inquiry, where every statement is to be backed up by scores of statistics, citations, calculations, and conclusions stated with tentative honesty: We are never sure, we can't really know. There are too many variables, we have only scratched the surface. When it comes to bird song, there is rarely enough data to guarantee bold generalizations about anything we hear.

The more I grapple with this wonderful field of bird song biology and ethology, the more it seems closer to artistic elegance than objective certainty. Not only are birds individuals, each one ready and willing to do exactly what you don't expect, but scientists are also individuals, and in their leaps of imagination they surprise even themselves. Peter Marler discovered that bird calls function more like a language than bird songs, with specific innate sounds of alarm and cries for food. Françoise Dowsett-Lemaire, with her ear to the thickets, decoded the pieces of the marsh warbler's panhemispheric litany of imitations, from Belgium through Tunisia down to the Zambezi. Wallace Craig heard the harmony of the universe in three notes from a pewee. These are leaps of listening, brilliant dreams. These scientists were diligent and meticulous but also creative. They invented beautiful pictures of the songs of birds. As with Nottebohm's discovery of neurogenesis in the canary, they all helped to illuminate a new vision of nature.

I started out believing that to make full human sense of the mysteries of bird song, we ought to use the whole toolbox of human talents. Poetry takes language to its limit to express the intensity of feelings and their bond to place. It is best at revealing the immediacy and necessity of bird sounds. Music knows that a series of sounds that have no message may convey beauty, forming the structure we hear as it goes. It presents the obvious rhythm and shape that bird songs show. And science wants to be true, impervious, and powerful because it must be far more than any one human's opinion about how the world looks or sounds. It graphs bird songs in a form rigorous enough for us to pick apart, like that half-second lyrebird note printed above. Sound immediately becomes more than anyone can hear, as technology leads us to trust our tools of analysis more than our senses.

The science of bird song, like the art of bird song, is rich in hypotheses but thin on conclusions. There are many good reasons why bird song *might* have evolved, what functions it *could* serve in the midst of mating and fighting. Current theory is buoyed by the arbitrary whims of sexual selection as a kind of random preference for peculiar traits. It wants to measure the degree of complexity in a song, the number of motifs, the sheer length, the amount of variation, all tabulated, an asymptotic curve flattening out to a limit. No more new songs, that's all I've got: Three for the pewee, twenty for the song sparrow, fifty for the thrush, one hundred for the nightingale, two hundred for the mockingbird, two thousand for the thrasher. Thresh, thresh, thresh, can anyone count so high without being moved by the relations of one sound to the next? Forty seconds for Webern, three minutes for your average pop song, twenty minutes for Mozart, one hour for Beethoven. Birds do not watch the clock.

This many-sided approach to bird songs is part of a greater project of finding ways the sciences and the arts might constructively come together. No scientist has tried harder to do this than Edward O. Wilson, the great entomologist, conservationist, and nature writer. His recent book *Consilience* proposes that all human knowledge ought to be unified upon the solid ground of science, which will in time come to illuminate and clarify all of our more far-flung creative endeavors.

He writes quite lyrically of science's need to reduce complexity down to its simplest parts in order to make sense of it. Reductionism remains science's most mundane and dominant tool, yet Wilson describes it poetically. "Let your mind travel around the system. Pose an interesting question about it. . . . Become obsessed with the system. Love the details, the feel of all of them, for their own sake. Design the experiment so that no matter what the result, the answer to the question will be convincing."

Yet, Wilson says, the love of complexity *without* reductionism is the path toward art, which illuminates the richness of nature with sheer celebration. Art, he says, transmits feeling, while science wants to figure it out. Wilson does not want science to be art or art to be science, but he does believe that science will soon be able to explain why we find art compelling. He imagines that one day we will be able to read brain activity accurately enough to know exactly what art is doing to us. The pioneer work now being done on bird brains may show the way. Science may not touch us emotionally, but only its rigorous method will be able to find the reasons for our deepest emotions. Like birds, we are born with part of our response and we must learn the rest. The human brain "constantly searches for meaning, for connections between objects and qualities that cross-cut the senses and provide information about eternal existence. . . . In order to grasp the human condition, both genes and culture must be understood, not separately in the traditional manner of science and the humanities, but together, in recognition of the realities of human evolution."

The best art, Wilson presumes, is that which is truest to our biological origins. Wilson guesses that art evolved to enable our sense of self-reflection, unique in the animal world, to bring us back into joyous contact with where we came from. Art's quality should be measured by its fidelity to human nature, its ability to reveal our essence and our species' place in nature. Necessity is the goal of invention and beauty in art should also be its truth.

There is something circular about such an argument. Few artists or art lovers would agree with it. Only after there is agreement that an artwork is good do we look for reasons for its goodness. Yet the hunt for authenticity does resonate with the reason why I believe birds sing. Music expresses the bird's essence just as it expresses ours. We are one species that is sometimes able to listen beyond ourselves and our needs. Beethoven and Mozart thought birds' musicality was a kind of a joke, so they played around with their songs in lighthearted ways. Messiaen, in contrast, took the real wildness of bird song most seriously, while personalizing it by blending in other influences—Greek and Hindu rhythms, and a unique system of harmony built out of dissonance. Pestel and I want to stay out there with the birds, play in their forests, changing the boundaries between what is music and what is not.

Not all humanists are comfortable with Wilson's wish that science will be able to prove the value of our aesthetics and culture. Poet and farmer Wendell Berry wrote *Life Is a Miracle* to specifically challenge what he saw as Wilson's elevation of reduction and explanation over wonder and experience. "We always know," writes Berry, "more than we can say. . . . There is no reason to assume that the languages of science are less limited than other languages." Why not instead urge scientists to take more seriously their *own* participation in the world of animality, affection, joy, and grief, that precede all our elucidations? Once after a lecture I asked Wilson if he really thought nature was a machine. "Oh no," he responded. "We have much better metaphors than that. But they are too complicated for the general public."

Any specialist has the tendency to believe his field contains all others. I learned in school that philosophy was the "queen of the sciences." Wilson wants science to explain everything. Berry wonders if both art and science can subordinate themselves to concerns larger than their own professional standards. Now standing at dawn in the Queensland rainforest, I imagine bird song contains all the paradoxes of any human contemplation of nature.

Art cannot be replaced by any analysis of its power, even if it comes straight from a natural pathway millions of years long. The song system in a canary's brain is no substitute for the song. It doesn't teach us to sing like a bird—Ayanna Alexander still practices on a disembodied syrinx day after day, seeking data that might teach her what it feels like to sing way beyond human breath and control. As Berry sees it, much of the power of life is not meant to be made clear. "I don't think creatures can be explained. I don't think lives can be explained. What we know about creatures and lives must be pictured or told or sung or danced."

Science is supposed to be different from art, but when it comes to bird song its conclusions seem strangely poetic. There are no answers at the end of bird song papers, just further dreams and guesses as to the elusive reasons why. It is no surprise that scientists at the summing-up stage of their work often turn to poetry, from Garstang's light verse to Skutch's yearning for a universal harmony that might hold the whole wild universe together. The great whale scientist Roger Payne spent one afternoon listening to Messiaen, humpback whale songs, and bird songs, and this led him into a swirling reverie in which sound, structure, and emotion coalesced. "Is it possible that the universe sings?" this scientist was not afraid to write. "Is it possible that God is the song of the universe?"

Before we are artists or scientists we are human beings, and when we confront phenomena as enigmatic and tenacious as bird song, no single one of our faculties is enough. The beautiful songs of life are older than our entire species, and they will continue long after all human music has dissolved. If the works of God are to be heard on Earth, there is no better place to find them than in the deep intricacies of incomprehensible bird song. If you'd rather leave God out of it, go ahead—the snail's pace of evolution itself may have been enough to make all this useless beauty, and all our need to imagine an intelligence behind the whole thing as well. I'm still using everything I learn and believe in order to listen. I desperately want to join in with the birds long before I have any idea what they are doing or why.

Science, poetry, and music say different things about bird songs, but in their quest for an ultimate answer to the question *why?* they begin to blur. John Clare used "*Wew-wew wew-wew chur-chur chur-chur*" to define his entire vision of poetic rhythm. Science didn't top this until Darwin offered a vision to link the interconnected noises of nature to purpose and desire. Even music had to reach a certain level of openness to the value of all sounds before the nightingale's rhythms could be appreciated as music.

So bird song challenges science and art alike to extend their reach. We are still grasping. It's still possible that bird song will turn out, like the dance of the honeybee, to contain all kinds of specific information that we will one day decipher, although given what we presently know, this is highly unlikely. Though the lyrebird's song is composed out of the sounds of other birds, it makes sense only as a whole, sung in this forest, at exactly the correct time, for reasons long ago subsumed into an unquenchable will to go on. It exhausts all attempts to describe it, like all the greatest music. Poets, though, will never stop trying. I know of no great lyrebird poetry, but George Meredith's late Victorian "Lark Ascending" does come to mind:

> Shrill, irreflective, unrestrain'd,
> Rapt, ringing, on the jet sustain'd
> Without a break, without a fall,
> Sweet-silvery, sheer lyrical,
> Perennial, quavering up the chord
> Like myriad dews of sunny sward
> That trembling into fulness shine,
> And sparkle dropping argentine. . . .
> We want the key of his wild note
> Of truthful in a tuneful throat,
> The song seraphically free
> Of taint of personality,
> So pure that it salutes the suns
> The voice of one for millions,
> In whom the millions rejoice
> For giving their one spirit voice.

That's the sort of verse explorers might have had in mind when they named this bird after their regent, Prince Albert, who never got to see or hear one himself. Meredith understood the bird's desires, writing elsewhere in the poem that he "craves nothing save the song." Music is a songbird's utmost desire, an endless yearning to sing.

"Wake up, David, George is waiting for us." Pestel goads me out of my reverie back into the present in the flickering rainforest, as the sun rises higher through the trees. A red-necked pademelon, kind of a small kangaroo, crashes off through the bush on my right. George has run to another vine platform. He begins again.

Once an Albert's lyrebird begins his cycle of mimicked sounds he cannot easily stop. His brilliant tail feathers are arched over his head, he rocks back and forth on the vines, making the whole forest shake. Why does he sing the same cycle of sounds on and on through the winter breeding season, month after month until his bright feathers fall off and he becomes an inconspicuous, wandering forager for the rest of the year? The mantra, writes Frits Staal, must be performed correctly.

The lyrebird is born to learn and it takes him half a decade to perfect his song. He doesn't need to know the *names* of all the birds he is imitating, and as a musician in his presence I don't need to know either. Listening is forgetting the names of the sounds you hear. This is not to belittle information or the immense value of scrutiny and detail, but when it comes to suddenly improvising a place in some new music, you need to erase all expectation while building on all your stored experience. While the bird sings, you must become a bird. George cocks his head, turns toward us, shakes a stick with his claws. It is time for the humans to come in.

Boo. Toot. Pe-bum, Brealummph! Our music does not proceed in such strange words but with melodies that are birdlike only by association. George at first is puzzled with the strange sounds. He pauses his concert for just a half second, but not much longer. "What are these strange foreign sounds getting in my way?" he might be thinking. "What right have they to make noise at this hour at this time of year? This is my time, none other, that's the way this world is designed."

This rainforest has a wonderful echo, not crisp like a concert hall but delicate as only a room muffled by living green leaves can be. *Whoooodleadleap. Burdelealap.* The territorial call of a Michael Pestel. I'd know that flute riff anywhere. I concentrate on the power of a single tone, high B. *Ping. Ping.* Some tiny forest bird above matches it. George cannot stop, but he can change his song—in the smallest, subtle ways in response to what he hears.

This is far more than we expected. I'm twenty feet away from the one Albert's lyrebird in the whole world who will let humans get close enough to watch him do his marvelous dance and song. What do I do? I'm not even respectful enough to stand quietly and appreciate the opportunity. I cannot resist playing along. My single notes soon extrapolate to phrases, jumps up and down. Imitations of imitations, mimicry of the mimic. I know, as usual, I'm playing too much to earn a place in this forest. I try to learn from the proportions of George's music—to quiet down, to hone my style.

The disconnected phrases, the copy of the copies. Tiny pieces here and there of all sounds heard and learned. This is the music that works for us, this is the way our lives are prescribed. Alec Chisholm, one of the early Australian writers on the lyrebird, also wondered why it is so rare to see a female anywhere near the site of a song display. Years of observing superb lyrebirds in parks close to Sydney convinced him that the bird does it for "sheer joy in life." How else could the male lyrebird sing with such inevitability and imperviousness to all intrusion? Performance poet Chris Mann, a man who owns a farm full of superb lyrebirds in Victoria who swear in Polish (a legacy from the previous owner), believes lyrebird song is more about politics than aesthetics. "It's practical, not beautiful. They define their community life through their music." He compares the bird's life to the original life of native peoples everywhere. "They spent more time singing and dancing than working. How could you say that the severe work ethic of English colonists was an advancement over that?"

Swedish scientist Björn Merker has studied duetting gibbons and the protomusical babble of children, and performed in techno rave bands doing multiphonic Mongolian chanting. He warns me to "steer clear of that nonsense of music expressing emotions. The only emotion music expresses is the *urge to sing*." If you want to be a lyrebird, you've got to learn the right song. It cannot be faked. "The song became the right song when the species was defined," says Merker. "That's when it matters. Now you are of that species, you do just what being a lyrebird requires. It is your destiny." As a human messing with this plan, I'm only tempting fate. The bird has no *need* to listen or toy with me. So what does it mean if he does?

From dawn until noon there are no female lyrebirds around. The other males in the distance respond most clearly to the specific short territorial call that began the session and is occasionally interspersed. But the whole mimicry cycle? George sings it because he loves it, because he must. Selection over eons does not require but merely allows this ultimate avian performance. George is a lyrebird possessed, overtaken by hormones, design, circumstances, season, whatever explanation you choose. He is adrift in a song of purpose that he cannot ignore.

At a certain moment in the cycle of cycles he moves on to the booming conclusion—the series of harsh sounds known as the *gronk* that accompanies a claw-by-claw new rhythm that shakes the sticks of his platform of vines swinging up on the sides to shake the booyong trees above. *Gruhnk gruhnk, tkdtdhklttdktdhk, gruhnk, gruhnk;* Kurt Schwitters would have especially loved these sounds had he the good fortune to come here. *Gruhnk gruhnk fzzz fzzzz fzzz fzzz,* a soft, almost electrical noise comes from George's syrinx. He shakes his delicate shimmering tail feathers in another tilted arc and freezes, just at the moment to catch the greatest power of the gleaming winter light.

The whole Australian bush sparkles as the wind blows. The tiny bright dots of Aboriginal art mirror this same twinkle as the dancing lyrebird, pulsing at the moment with all of nature's visual energy. Contemporary Aboriginal musicians such as Herb Patten imitate a range of bird sounds by blowing across a gum leaf, becoming like lyrebirds themselves. Native Australians called the lyrebird *bulen bulen,* a word that sounds suspiciously like *bulbul,* the Persian name for the nightingale. Aborigines have a story about the lyrebird giving all animals their different voices, cutting through the babel of confusion just like the Hopi mockingbird. All over the world different bird songs gird the same myths.

The *gronking* is supposed to signal any available female that the time for mating has come. But George's chance of luring one in are small. Most of the females are down in the gullies, closer to the more wary male lyrebirds, and George's territory is higher up, on a hillside. It's good feeding land, but it can be lonely. Glen Threlfo has heard George pause from his song to spew out a mass of angry animal sounds rarely heard: shrieking possums, fighting pademelons, general squeals. "Right," he remembers, "I reckon it's all sheer frustration."

Few people have ever seen an Albert's lyrebird in the midst of mating. Syd recalls that only once, in his thirty-odd years of studying the sounds of this elusive bird, he decided to record all the song output of a single bird calling from his perch of low-slung vines on Tamborine Mountain. The bird began to repeat the precise string of mimicry, one time after another, this stream of sounds set for all members of the tribe. Then after a time he heard the telltale rhythmic *gronks* that signal the male is ready for the female to cut to the chase. All of a sudden Syd heard a rustle, turned around, and there was the female lyrebird heading straight toward the male through the underbrush. The problem was, Syd sat in the middle of the path, right in the way from girl to boy. She had never seen a creature like Syd and had no idea what to do. She must get closer to the male, and she was unprepared for this strange obstacle and his extreme decorations of microphones and wires and headphones, rivaling the outlandish feathers of the male lyrebird in sheer extravagance. So what did she do?

Female lyrebirds rarely sing, but like many other songbird species they can do it if they have to. And this one had to get some word to the male. She started the Albert cycle herself. At first it was raspy. She never practiced much. The bowerbird call squeaked. The rosella wasn't quite right. But soon she stepped into it, she's got it in her, she'd heard this sound from males all her long life. Soon it came out clear as day, she knew the song just fine, but until that moment when an ornithologist blocked the path to her mate there was never a reason for this bird to sing. In this rather sexist lyrebird tale, it turns out females *can* sing—if driven to it.

Glen has seen George mating only once. "His tail was up in the air, just like those old paintings everyone says are incorrect." Only in the heat of passion is the lyrebird's tail held up in the shape of a lyre. "Ah," Glen smiled, "he really couldn't help himself."

"Maybe this bird knows that the yearning for the act means much more than the thing in itself," Pestel suggests. "It certainly takes a lot more of his time."

We hear George displaying from another platform and approach him rather rapidly through the brush. Just past the bowerbird bower, littered with blue plastic spoons, he turns sideways, flicks his tail over his head, and utters that whole series of alarm cries and possum screeches Glen warned us about. Maybe George's frustration concerns us, who are busy interrupting his solo concert, and he trots off to the north edge of the territory, past the human trail, off to forage a bit more. We hear nothing from him for some time. The *whoom whoom whoom* of a distant ground pigeon suddenly gives the soundscape a beat.

For an hour we tune in to the smaller, twittery birds of the canopy, all making subtle interacting sounds as part of some rhythmic, ecological symphony. I think of Messiaen and his attempt to reproduce the whole avian milieu of a place in a series of piano pieces. What audacity! The rooted flow of nature establishes the music's structure. The start and stop of this ingrained pattern determines the actual rhythm, not some steady pulse decreed from above.

Wait! Down the hill . . . at those display vines just above the trail to Pademelon Rock, the very edge of his territory . . . he's back. We run silently through the forest. The display is on. The *braaamph* of the rosella and *wheeooowheep* of the bowerbird combine precisely in George's cycling song. There are four large green catbirds above us in the trees. They're the only ones who can see all that's happening. One stares straight at me as I put the clarinet to my lips. Then he cocks his head toward Michael, two flutes up his nose. Then at George. *This* they haven't seen before—they *meow* up to others in the trees. The catbirds all arrange themselves for the best view.

I listen to one full cycle of George before I join in. It is my wish to play music worthy of this bird's acceptance. Like the Kaluli performers in New Guinea, awash in feathers, dance, and song, I too would become the bird. A great lyre of imaginary feathers girds my sound, curving over to surround my mind with crucial music. To you they are birds, to me they are musicians, their songs ever the same, while we hunt for our proper place among their powerful voices in the forest.

A few short sharp notes, a mix of the squawk and the pure. My own chromatic slightly speeding-up version of his descending territorial song ending in an octave leap: A' G# G F# F E D# A A'! Not sure if George hears this as anything akin to his own announcement and defense. You are supposed to copy birds exactly if you are to convince them you're doing the same as they are, since they're supposed to hear absolute pitch, not relative. But that is a playback experiment. With interspecies music it is better not to copy but to play and build. Like Messiaen with his birds as guides, not models, I am trying to play in and around George, the one Albert's lyrebird willing to face human music on his own ground.

In this sonogram George is in black and I am in gray. "I do believe you're getting to him," muses Syd, listening closely to our proceedings, wool-muff headphones on his ears, his own digital recorder

INTERSPECIES MUSIC AT LAST

plugged into a giant parabolic reflector. "I've been recording George for twenty years and his song is different today. He sticks with the parrot squawks and bowerbird sweeps, only rarely going on to the kookaburra, honeyeater, and robin notes. He begins the cycle over and over, but seems unable to finish it. Never done that before. Either he's changing the elements of his routine or you two strange foreign birds are doing something to his demeanor."

The song continues, I try to place my clarinet in and around the breaks. I am dancing in the underbrush, crackling branches underfoot, leaping up and down, curving my shoulders in imitation of those long delicate feathers I'll never swing. The resonance of the clarinet in the forest is full and clear, as if the wood that made the horn has somehow returned to its place. I am an interloper, unlike George. My music is not rooted to this place, I am a migrant, a wanderer, a marsh warbler who picks up songs wherever he goes.

George, he never goes far. His songs must be performed exactly on the vines. This is site-specific music. The winter forest needs his song to carry far through the dense leaves, so other birds can hear him a thousand meters away. "He's the greatest performer in the animal world," Glen announces with pride. "The sheer volume of the song, the perfect mimicry. And on top of all that he does a dance. Compared to George, a peacock's just a deaf-mute." He took five years to make his film of George, *The Albert Lyrebird: Prince of the Rainforest,* and he still gazes at his subject with awe and admiration.

This play-along experiment cannot be simply analyzed with no variables or controls to be isolated for the test. Music is first hunted for and then constructed out of wise listening and the hope that the birds and the leaves will let us in. The result is not a table or a conclusion but a human music that is radically changed by the close influence and presence of birds. I've learned these imitations and territorial songs not by their function but by their musicality. They are the components of the lyrebird's song that aurally defines this forest, an emblem of the acoustic heritage of what may be (if we are to believe a team of Australian and Swedish paleobiologists) the oldest singing bird on the planet. Nearly a hundred million years ago songbirds emanated across the globe from Gondwanaland, with a generalized superbird like the lyrebird a probable ancestor. He perches, he glides, he sings, he dances, his plumes are magnificient. His leaping silhouette at dawn resembles the famed archeopteryx who is the link between dinosaur and bird, the singing ghost of a transitory past living on into our time.

As the great tail flips down and George disappears into the wood, Pestel is transfixed. "I may never be able to play along with a mere sparrow again."

BACK HOME I try to listen again to human music, but it doesn't seem right. I've spent too long with the lyrebirds, and I've begun to hear music their way. The world's bird songs are not infinite, but there is so much more we can learn from them. Species are supposed to sing only for their own kind, but the more I listen, the less I'm sure. Yesterday I heard all the thrushes singing together in the laurel woods: veeries, hermits, wood thrushes. They all seemed to be getting at one total song.

You don't have to forget the names of the birds you hear, but you should not think that simply identifying them by their voices is enough. We may listen with love as we enumerate. The songs of birds make our whole world more vibrant and alive. Rachel Carson's warning of a "silent spring" brought enough fear to begin the worldwide environmental movement forty years ago. We have since learned to care, and our springtimes resound with immeasurable beauty. Are bird songs now in danger? Is the restless noise of humanity pushing them away?

The science is, of course, inconclusive. Sure, forests are being cut down in North American breeding grounds, and in tropical wintering areas as well. Birds that require large unbroken woodlands to prosper are finding their numbers in decline. There are half as many wood thrushes as there were forty years ago, but there are still fourteen million of them. Species that thrive in small open tracts and tiny forest patches and lawns are doing better than ever. In precolonial times the American robin was fairly scarce. Now it is the most common bird heard in suburban backyards. But still *no one* has bothered to study its song. Here's a project any willing listener can commence with, they are singing everywhere. Start to hear them yourself, and ideas will come. All robins have a lilting, singsongy identifiable sound, but each bird has variations. How? Why? Listen and you may find out. Play along and you may become a bit of a robin yourself.

Introduced species such as the starling and house sparrow, both originally from England, have pushed out native bluebirds and purple martins, and these usurpers are reviled by bird watchers and others who prefer their environments native and pristine. You may find starlings ubiquitous and annoying, but remember: their songs are at the very edge of our ability to grasp, one of the clearest examples of the idea that different bird species can have a particular aesthetic sense—"Way down upon the Swa . . ."

I do not want to encourage complacency in this journey toward greater wonder at the music of birds. Nature is certainly under siege, and the more attention we pay it, the more likely we may be able to save humanity's place in the larger enveloping world. E. O. Wilson, however much he wants science to gird our sense of culture, looks to ethics when he wants to find a reason for humanity to rally for the cause. Preserving biodiversity is simply a good. Species disappear daily at an alarming rate. Rare songs become ever rarer. Birds about which nearly nothing is known disappear before we take the time to listen to them. We will come to realize forests and other essential habitats must be protected, Wilson writes, not because we need more things to study but because it is the right thing to do. Science needs morality if it is going to save the world. The Kaluli are right: birds are the voices of the forest. They can tell us more than we will ever be able to know.

Why do birds sing? For the same reasons we sing—because we can. Because we love to inhabit the pure realms of sound. Because we must sing—it's the way we have been designed to tap into the pure shapes of sound. We celebrate this ability in our greatest tasks, defining ourselves, defending our places, calling out to the ones we love. But form remains far more than function. We spend lifetimes immersed in the richness of these creations. "Figure out" a symphony or a nightingale strophe and both sounds will still need to be performed, on and on, way past the time in which your answer will be adequate. No explanation will ever erase the eternal need for song.

The old becomes the new, as these avian musics test the limits of our human need to sing and play. Our rituals may be based on them, our melodies derived from them, but we can still be challenged to bring our music closer to nature's slowly written phrases, whose metrics have evolved over many more years than any human could practice or refine. "You're still serious about playing along with birds?" people ask me. And I answer that it's the most serious music I can imagine, the deepest and most eternal. "So why do you try to compete with it?" they say. Defending my territory, I too may seek admirers, but mostly I want to come closer to life's most essential timeless sounds. After too many years, maybe only the birds will play along with me, my rhythms and pitches veeried far from human strains, way beyond reason and function.

No matter what I learn and how little I know, I will never give up the chance to make music together with birds. To wing it, so to speak, and wait for what will *cheep* in return. Like all art, bird song works best when we let it play on. Like science, it is built on the music of endless previous generations, still evolving into new sounds. The music made the questions begin, but no answer will erase the gift of the song, one simple offering from human to animal and back.

Acknowledgments

THANKS TO ALL the scientists and musicians in the book for agreeing to talk with me, especially Peter Marler, Fernando Nottebohm, Partha Mitra, Françoise Dowsett-Lemaire, Erich Jarvis, Douglas Quin, Magnus Robb, and Björn Merker.

Thanks to my agent, Kathleen Anderson, for helping to shape this project from the beginning, and to acquiring editor Amanda Cook, and executive editor William Frucht at Basic and Jon Turney at Penguin for their sage advice. Special thanks to Wandee Pryor for reading the first draft so closely and helping to smooth it.

Later drafts were expertly read by John Horgan, Eric Salzman, Evan Eisenberg, and Joan Maloof. The neuroscience chapter was scrutinized by Linda Wilbrecht. They pointed out many wrong turns and obvious mistakes, and those that remain reflect my stubborn wish to stand my ground.

Thanks to Rhonda Greene at the interlibrary loan desk at the Van Houten Library at the New Jersey Institute of Technology, who endured my annoying and sometimes redundant requests for obscure bird song tomes in various languages. You got them all.

Thanks to Michael Pestel for years of cross-species musical encounters, beginning in the Pittsburgh Aviary and continuing down under. In Australia, Kate Rigby, Vicki Powys, Robin Ryan, Gisela Kaplan, Barry Craig, Jan Incoll, Alex Maisey, and unflagging Syd

Curtis made our trip a smashing success. We and the lyrebirds will never be the same again.

Other readers and helpers over many years of my investigations of more than human music include David Abram, Peder Anker, Tim Birkhead, Chip Blake, Tim Blunk, Patricia Cleveland-Peck, John Coakley, Jim Cummings, Rebecca Danielsson-DeRoche, Emily Doolittle, David Dunn, Lang Elliot, Aundrea Fares, Jay Griffiths, David Hindley, Sabine Hrechdakian, Gayle Johnson, Fred Jüssi, Jay Kappraff, Carol Krumhansl, Majid Labbaf, Matthew Leonard, Annea Lockwood, David Lumsdaine, Dario Martinelli, Steve Mercier, Jim Metzner, Jim Nollman, Richard Nunns, Geoffrey O'Brien, John P. O'Grady, Katy Payne, Richard Powers, David Robertson, Daniel Rothenberg, Ben-Ami Sharfstein, Grant Sonnex, Frits Staal, Charlotte Strick, Allan Thomas, David Toop, Marta Ulvaeus, Rene van Peer, Maya Ward, Lawrence Weschler, and Lisa Westberg.

Thanks to my son Umru for learning to sing like a bird before speaking like a human, and to my wife Jaanika for putting up with my increasing obsession with the enigmatic sounds of other species. Thanks to my father for believing so much in this project and encouraging me throughout my sensitive learning period.

Further Reading

Why Birds Sing tells a story that traverses the science, music, and poetry of bird song, so there are many directions you can go for further exploration. Check the endnotes for references to specific scientific articles and other academic sources, but here are some of the best books in the several fields I have tried to cover.

In science every ten years or so a new compendium of all that is known at the time on bird song comes out. The latest and by far the most comprehensive is Peter Marler and Hans Slabbekoorn's *Nature's Music: The Science of Birdsong* (London: Elsevier, 2004), an anthology with contributions by many of the scientists discussed herein, complete with audio CD examples. Also up to date is the more technical collection edited by Phillip Zeigler and Peter Marler, "Behavioral Neurobiology of Birdsong," *Annals of the New York Academy of Sciences* 1016 (2004). A revised version of the same volume will be published by Cambridge University Press in late 2005. Together these should give you more than you need to know of where science currently stands on the question of why and how birds sing.

Other notable scientific surveys over the years include Clive Catchpole and Peter Slater, *Bird Song: Biological Themes and Variations* (Cambridge: Cambridge University Press, 1995); Donald Kroodsma and Edward Miller, eds., *Ecology and Evolution of Acoustic Communication in Birds* (Ithaca: Cornell University Press, 1996); Donald Kroodsma and Edward Miller, eds., *Acoustic Communication in Birds*, 2 vols. (New York: Academic Press, 1982); Rosemary Jellis, *Bird Sounds and Their Meaning* (Ithaca: Cornell University Press, 1977); R. A Hinde, ed., *Bird Vocalizations* (Cambridge: Cambridge

University Press, 1969); and still perhaps the finest survey volume (of course quite dated) on the behavioral questions surrounding bird song, Edward Armstrong, *A Study of Bird Song* (London: Oxford University Press, 1963; Dover reprint, 1973).

On the subject of literary references to bird song you will find plenty of examples in the collected poetry of John Clare and Walt Whitman, and there are several fine anthologies of birds in literature at large: Edward Armstrong's *The Life and Lore of the Bird in Nature, Art, Myth and Literature* (New York: Crown, 1975) is a fine compendium of words and images. Jen Hill, ed., *An Exhilaration of Wings: The Literature of Birdwatching* (New York: Viking, 1999) is a suggestive collection of short quotes from the 1600s to the early 1900s. Dylan Nelson and Kent Nelson, eds., *Bird in the Hand: Fiction and Poetry About Birds* (New York: North Point Press, 2004) is an excellent gathering of contemporary works. Lang Elliot's wonderful coffee table book, *Music of the Birds: A Celebration of Bird Song* (Boston: Houghton Mifflin, 1999) has exuberant photographs, a discerning essay linking poetry and science, and an impeccable CD of seventy-six wondrous songs. Of course Walter Garstang's *Songs of the Birds* (London: Bodley Head, 1923) is in a poetic genre all its own.

When it comes to music, the literature is surprisingly thin. Charles Hartshorne's *Born to Sing: An Interpretation and World Survey of Bird Song* (Bloomington: University of Indiana Press, 1973) is the most philosophical approach to the subject, in a class by itself. On the possible musicality of nature see Dario Martinelli, *How Musical Is a Whale? Towards a Theory of Zoomusicology* (Imatra: Semiotic Society of Finland, 2002)—see www.umweb.org/dm/act.htm for details on its availability; and the large anthology based on human evolution edited by Nils Wallin, Björn Merker, and Stephen Brown, *The Origins of Music* (Cambridge: MIT Press, 2000).

For a compendium on how musicians have viewed the relevance of nature in general see my own anthology, David Rothenberg and Marta Ulvaeus, eds., *The Book of Music and Nature* (Middletown: Wesleyan University Press, 2001), and David Rothenberg, *Sudden Music: Improvisation, Sound, Nature* (Athens: University of Georgia Press, 2002), for my own more personal views on the subject. In terms of transcribing bird song as music, F. Schuyler Mathews, *Field Book of Wild Birds and Their Music* (New York: Putnam, 1921) is the classic as described in the text. It is reprinted every few years, although the older editions are more complete. Eighty years later, Messiaen's two six-hundred page volumes of detailed bird song transcriptions are an incredible resource few people have looked at: Olivier Messiaen, *Traité de rythme, de couleur et d'ornithologie,* vol. 5, books 1 (Europe) and 2 (The Rest of the World) (Paris: Alphonse Leduc, 2000). The music of most of the composers I mention in Chapter 9 can be easily found on CDs.

Historically it is important to understand the role of evolution in changing our understanding of natural aesthetics. Charles Darwin, *The Descent of Man* (1871) has a surprising amount of material on birds, as well as an often overlooked companion book, Charles Darwin, *The Expression of Emotions in Man and Animals* (1872). The best modern accessible history of sexual selection is Helena Cronin, *The Ant and the Peacock* (Cambridge: Cambridge University Press, 1991).

The little-known and beautiful book from Darwin's time by Jules Michelet, *The Bird* (London: Wildwood House, 1981 [1879]), shows a French slant on rhapsodic nature writing. In addition to the classic writings of John Burroughs, especially *Birds and Poets* (Boston: Houghton Mifflin, 1904), important late nineteenth-century works of bird lore include Wilson Flagg, *A Year with the Birds* (Boston: Estes and Lauriat, 1881); Bradford Torrey, *Birds in the Bush* (Boston: Houghton Mifflin, 1893); and Simeon Pease Cheney, *Wood Notes Wild* (Boston: Lee and Shepard, 1891), which is especially useful because it contains a compendium of previous transcriptions and conclusions on the meaning of bird song.

The early twentieth-century bridge between the naturalist and scientific approaches can be found in Elliot Howard, *Territory in Bird Life* (London: William Collins, 1920); and, of course, Margaret Morse Nice, *Studies in the Life History of the Song Sparrow,* 2 vols. (New York: Dover, 1964 [1937]), the most detailed published report of a long-term observation study of any bird species. The finest recent bird essays in the naturalist tradition are Alexander Skutch, *The Minds of Birds* (College Station: Texas A&M University Press, 1996); and Calvin Simonds, *Private Lives of Garden Birds* (North Adams: Storey Press, 2002).

It should be noted that in contrast to the vast scientific literature there is a counterliterature that shuns generalization, focusing instead on human experiences with particular animals. Len Howard, *Birds as Individuals* (Garden City: Doubleday, 1953) is the classic of this genre. Every bird is different and must be studied and listened to as such. Viscount Grey of Fallodon in *The Charm of Birds* (New York: Frederick Stokes, 1927) went so far as to assert on page one that his book "will have no scientific value." The sample size of individual experience is far too small. There is one scientist, however, who has spent eight years studying and documenting her work with one single bird: Irene Pepperberg, *The Alex Studies: Cognitive and Communicative Abilities of Grey Parrots* (Cambridge: Harvard University Press, 2000). So science has found a way to admit the individual bird after all.

There are several important books on the individual species discussed in the text. On the most famous of European songsters see Richard Mabey, *The Book of Nightingales* (London: Sinclair-Stevenson, 1997). You know by now there's a two-hundred-page book on a three-note song: Wallace Craig, *The Song of the Wood Pewee: A Study of Bird Music,* New York State Museum Bulletin, no. 334 (Albany: June 1943). Copies of this bound report can sometimes be found at: www.abe.com. On lyrebirds you'll have to go to Australia, and even there you won't find a single book on the amazing beast in print, but a few can be found in libraries. L. H. Smith, *The Life of the Lyrebird* (Richmond, Victoria: W. Heinemann, 1988) is the best, although I hear Glen and Syd are each working on their own books about George. There are no good books exclusively on mockingbirds or marsh warblers, at least not yet.

To hear many of the sounds and to see a video of the clarinet and laughing thrush duet, go to www.whybirdssing.com.

Notes

The six engravings of birds important to this story are from A. E. Brehm, *Foglarnes Liv,* translated from German into Swedish by J. E. Wahlström (Stockholm: Girons Forlag, 1875). Given the age of the volume, these illustrations may not be up to the standards of accuracy demanded by today's field guides, and in some cases they depict species slightly different from those described in the text. But they should give you the gist of these singers' spirit.

Preface

x seventy million people in the United States . . . call themselves bird watchers: An oft-quoted statistic from the National Audubon Society at: www.colszoo.org/news/beastban/bbnov03/feathered.htm

xii each whale has its own distinctive click train rhythm: Michel André and Cees Kamminga, "Rhythmic dimension in the echolocation click trains of sperm whales," *Journal of the Marine Biological Association of the United Kingdom* 82 (2000): 163–69.

Chapter 1

3 we'll never know what it's like to be a bat: Thomas Nagel, "What Is It Like to Be a Bat?" *Philosophical Review* 83, no. 4 (1974): 435–50.

4 listening, playing, composing: John Cage, *Silence* (Middletown: Wesleyan University Press, 1964), 15.

Chapter 2

14 *chup-chup-zeeee!:* For birdsong mnemonics, see www.elwas.org/birding/birdsongs2; www.geocities.com/Yosemite/2965/mnemonic.htm; or www.1000plus.com/BirdSong/birdsngv.html

15 "they heard the charming noise": Lucretius, "De Rarum Natura," translated by Creech, 1685.

16 "it's the bird of my loneliness": Kim Addonizio, "The Singing," in *Birds in the Hand*, ed. Dylan Nelson and Kent Nelson (New York: Farrar, Straus & Giroux, 2004), 142.

16 "to us, they are voices in the forest": Steven Feld, *Sound and Sentiment* (Philadelphia: University of Pennsylvania Press, 1990), 45.

17 the Temiar long for the world of spirits: Marina Roseman, *Healing Sounds from the Malaysian Rainforest* (Berkeley: University of California Press, 1993), 172.

17 The Hopi recognize deep information in bird song: H. R. Voth, *The Traditions of the Hopi,* Field Museum Anthropological Series, 1905. See www.earthbow.com/native/hopi/underworld.htm

17 "attainable border of the birds": David Guss, ed., *The Language of the Birds* (San Francisco: North Point Press, 1985), 202.

17 "Birds . . . *are* ideas": Paul Shepard, *The Others* (Washington: Island Press, 1997), 130.

19 "their voices are not destined to entertain human beings": Athanasius Kircher (1650), original text reprinted in Simeon Pease Cheney, *Wood Notes Wild* (Boston: Lee and Shepard, 1891), 218. Translation from Latin by Dario Martinelli.

20 "the third sound is called *recording*": Daines Barrington, quoted in *The Bird Fancyers Delight,* ed. Stanley Godman (London: Schott, 1954 [1717]), iv.

21 "food and music are administered at the same time": H. G. Adams, quoted in W. H. Thorpe, "Comments on 'Bird Fancyer's Delight,'" *Ibis* 97 (1955): 250.

23 "*wholly destitute of taste*": Immanuel Kant, *The Critique of Judgment*, trans. James Creed Meredith (Oxford: Clarendon Press, 1928), 89.

Chapter 3

34 "a taste for the beautiful": Charles Darwin, *The Descent of Man* (Chicago: Brittanica, 1952 [1871]), 451.

35 "sexual liaisons with hundreds of groupies": Geoffrey Miller, "Evolution of Human Music Through Sexual Selection," in *The Origins of Music,* ed. Nils Wallin et al. (Cambridge, MA: MIT Press, 2000), 331.

35 "animals utter musical notes": Charles Darwin, *The Expression of the Emotions in Man and Animals* (New York: Philosophical Library, 1955 [1872]), 87.

35 "the delight given by its melody": Ibid., 89–90.

36 "anger, triumph, or *mere happiness*": Darwin, *Descent of Man,* pt. 2, chap. 13, 457.

36 "the hideous music admired by most savages": Ibid., pt. 1, chap. 3, 302.

38 the preference for the trait with the trait itself: This introduction to R. A. Fisher's theory comes from Helena Cronin, *The Ant and the Peacock: Altruism and Sexual Selection from Darwin to Today* (New York: Cambridge University Press, 1991), one of the finest books on the historical difference between theories of sexual and natural selection.

39 females prefer the songs of dominant males: The various statistics of mating success correlated with excellence in song come from William Searcy and Malte Andersson, "Sexual Selection and the Evolution of Song," *Annual Review of Ecology and Systematics* 17 (1986): 507–33.

41 Music for the nightingale is not cheap: Robert Thomas, "The costs of singing in nightingales," *Animal Behaviour* 63 (2002): 959–66.

43 "The song is not all in the singing": John Burroughs, "Bird Songs," in *Ways of Nature* (1905), reprinted in *The Complete Writings of John Burroughs,* vol. 11 (New York: Wise, 1924), 35.

43 "Do they rejoice like the clouds": Wilson Flagg, *A Year with the Birds* (Boston: Estes & Lauriat, 1881), 214.

44 "All are regarded as love-notes": B. Placzeck quoted in Simeon Pease Cheney, *Wood Notes Wild* (Boston: Lee and Shepard, 1891), 139. This is a remarkable early book of transcriptions and essays on bird songs, and it is especially useful because it contains a compendium of earlier accounts of the subject, almost like an ancient web site.

44 Why indeed do you sing yourself?: Paraphrased from Bradford Torrey, *Birds in the Bush* (Boston: Houghton Mifflin, 1893), 33–34.

45 "we are invading a completely immaterial world of things": Walter Garstang, *Songs of the Birds,* 2d ed. (London: Bodley Head, 1923), 11.

46 "this transformation of the family rattle!": Ibid., 43.

46–47 "prolonged elevation of the spirit": Ibid., 36, 38.

48 "echoes from the wells of deep emotion": Ibid., 103–4.

51 "there is still no point to his song": F. Schuyler Mathews, *Field Book of Wild Birds and Their Music,* 2d ed. (New York: Putnam, 1921), xxii.

54 "Bird songs are most ethereal things": Ibid., 261.

55 flutelike elements in their song: Pauline Reilly, *The Lyrebird* (Kensington: New South Wales University Press, 1988), 47.

Chapter 4

59 "'you missed it again'": Ludwig Koch, *Memoirs of a Birdman* (London: Country Book Club, 1956), 106.

59–60 "I am a fanatic": Ibid., 180.

60 "I really began as an amateur bird watcher": Peter Marler, interview with author, Davis, CA, March 2004.

63 "almost magical potential": Ibid.

65 Sonograms . . . spit out by a computer: The most professional sound analysis software specifically for bird sounds is Raven, which comes from the Cornell Laboratory of Ornithology, at: www.birds.cornell.edu/brp/Raven/Raven.html. There are several free programs that produce fairly good sonograms. I have had good luck with Amadeus, at: www.hairersoft.com/Amadeus.html

66 "It's not like Watson and Crick": Peter Marler, interview with author, Davis, CA, March 2004.

66 "distinguishing bird from human language": W. H. Thorpe, *Bird-Song: The Biology of Vocal Communication and Expression in Birds* (Cambridge: Cambridge University Press, 1961), 11.

70 male and female birds perceive the same song a bit differently: Albertine Leitão and Katharina Riebel, "Are good ornaments bad armaments?" *Animal Behaviour* 66 (2003): 161–67.

70 "We are left to puzzle over the resulting richness": Clive Catchpole

and Peter Slater, *Bird Song: Biological Themes and Variations* (Cambridge: Cambridge University Press, 1983), 191. See also Peter Marler and Hans Slabbekoorn, eds., *Nature's Music: The Science of Birdsong* (London: Elsevier, 2004).

71 "bird songs represent music": W. H. Thorpe, *Animal Nature and Human Nature* (New York: Doubleday 1974), 307.

71 "its peak in Mozart": W. H. Thorpe, *Duetting and Antiphonal Song in Birds, Behaviour*. Supplement 18 (Leiden: E. J. Brill, 1972), 160.

72 "dead end in a scientific sense": Peter Marler, interview with author, Davis, CA, March 2004.

74 "Joan suffered in a way": Ibid.

78 how pure the song is: Eugene Morton and Jake Page, *Animal Talk* (New York: Random House, 1992), 179.

80 "It banishes all trivialness": Henry David Thoreau, *Journals*, 22 June 1853.

80 "'O spheral, spheral!'": John Burroughs, *Wake-Robin* (New York: Hurd and Houghton, 1871), 52.

81 "'That makes me feel queer'": Arthur Cleveland Bent, *Life Histories of Familiar North American Birds* (New York: Harper, 1960), 271. Online at: www.birdsbybent.netfirms.com/ch91-100/hermthrush.html

83 "Drip drop drip drop drop drop drop": T. S. Eliot, "The Waste Land," ll. 345–58.

84 "It is true micromusic": P. Szöke, W. H. Gunn, and M. Filip, "The Musical Microcosm of the Hermit Thrush," *Studia Musicologica Academiae Scientarum Hungaricae* 11 (1969): 431.

86 twice as good as we are: Robert Dooling et al., "Auditory Temporal Resolution in Birds," *Journal of the Acoustical Society of America* 112, no. 2 (2002): 748–59.

89 "like a veery, and no other: Daniel Weary, Robert Lemon, and Elizabeth Date, "Acoustic features used in song discrimination by the veery," *Ethology* 72 (1986): 199–213.

90 "literally out of practice": W. H. Thorpe and Joan Hall-Craggs, "Sound Production and Perception in Birds as Related to the General Principles of Pattern Perception," in *Growing Points in Ethology,* ed. Gregory Bateson and Robert Hinde (Cambridge: Cambridge University Press, 1976), 187.

90 somebody must discover the rules: Peter Galison, "Judgment Against Objectivity," in *Picturing Science, Producing Art,* ed. Caroline Jones and Peter Galison (New York: Routledge, 2001), 327–59.

90 "the brightest dreams of liberty!": Jules Michelet, *The Bird* (London: Nelson and Sons, 1879), 286.

Chapter 5

94 "My hearing isn't what it used to be": Françoise Dowsett-Lemaire, interviews with author, February–April 2004.

95 "the transition from 'mixed' to adult song": Françoise Dowsett-Lemaire, "The imitative range of the song of the marsh warbler *Acrocephalus palustris,*" *Ibis* 121 (1979): 453–68. See also Françoise Dowsett-Lemaire, "Vocal behaviour of the marsh warbler," *Le Gerfaut* 69 (1979): 475–502.

98 "They get fond of uttering particular words": Pliny, *Natural History,* vol. 10, ix.

98 "harmony with neighbouring sounds": Charles Witchell, *The Evolution of Bird-Song* (London: Adam and Charles Black, 1896), 229.

98 "chuckle, mock, and defy": Ibid., 220.

100 "a peculiar creaking quality": Marcel Eens, "Understanding the complex song of the European starling," *Advances in the Study of Behaviour* 26 (1997): 358.

103 "to digest the vocal bite": Meredith West, Andrew King, and Michael Goldstein, "Singing, socializing, and the music effect," in *Nature's Music: The Science of Bird Song,* ed. Peter Marler and Hans Slabbekoorn.

104 "a sequence of kisses": Marianne Engle and Meredith West, "Interspecies interaction: A tool for study of mimicry and species-typical birdsong," in *Crossing Interspecies Boundaries,* ed. D. L. Herzing (Philadelphia: Temple University Press, in press).

107 "the earth note should prevail": Samuel Harper, *Twelve Months with the Birds and Poets* (Chicago: Ralph Fletcher Seymour, 1917), 83–84.

107 "the great lies told with the eyes half-shut": Richard Wilbur, "Some Notes on 'Lying,'" in *The Catbird's Song: Prose Pieces 1963–1995* (New York: Harcourt, 1997), 137.

108 "they will begin to wildly improvise": Andrew King and Meredith West, "The effect of female cowbirds on vocal imitation and improvisation in males," *Journal of Comparative Psychology* 103 (1989): 39–44.

109 "He hastens on from phrase to phrase": Charles Harthorne, *Born to Sing* (Bloomington: University of Indiana Press, 1973), 123.

109 "Must not singing be enjoyable in itself?": Ibid., 54–55.

112 "*disliked* might be better": Calvin Simonds, *The Private Lives of Garden Birds* (Williamstown, MA: Storey Books, 2003), 18.

112 "Hello, hello, yes, yes, yes": Arthur Cleveland Bent, "The Brown Thrasher," in *Life Histories of North American Birds* (New York: Harper, 1960). Online at: www.birdsbybent.netfirms.com/ch31-40/thrasher.html

Chapter 6

118 "rarely repeated exactly": Margaret Morse Nice, *Studies in the Life History of the Song Sparrow,* vol. 2. (New York: Dover, 1964 [1937]), 121–22.

119 "an expression of excess energy": Ibid., 145.

119 "all moments of excitement": Ibid., 148.

120 He never returned to Interpont again: Some of this account of Nice borrows from Joseph Kastner, *A World of Watchers* (New York: Knopf, 1986), 153–54.

122 playing their own songs back . . . and watching what they did: Jeffrey Podos et al., "The organization of song repertoires in song sparrows," *Ethology* 90 (1992): 89–106.

123 nearly two hundred pages in length: Wallace Craig, "The Song of the Wood Pewee *Myochanes virens linnaeus:* A Study of Bird Music," *New York State Museum Bulletin,* no. 334 (June 1943).

123 the average song contains 750 phrases: Ibid., 175.

126 overlapping cycles of behavior: Wallace Craig, "Appetites and Aversions as Constituents of Instincts," *Biological Bulletin* 34 (1918): 91–107.

127 "Art *is* a fact": Craig, "Song of the Wood Pewee," 161–62.

130 tones zooming by: Olavi Sotavalta, "The flight-sounds of insects," *Acoustic Behaviour of Animals,* ed. R. G. Busnel (Amsterdam: Elsevier 1963), 374–90.

130 "the rattle of a tambourine": Olavi Sotavalta, "Song patterns of two Sprosser nightingales," *Annals of the Finnish Zoological Society "Vanamo"* 17, no. 4 (1956): 5.

133 birdsong . . . a form of *zikr:* John Baily, "Afghan Perceptions of Birdsong," *The World of Music* 39, no. 2 (1997): 51–59.

134 they got out their tabla drums . . . to jam along with the tape: You can hear this bit of Asian interspecies music and read the whole story at: www.open.ac.uk/Arts/music/mscd/baily.html

135 "Song . . . must be more than war": Mahdi Noormohammadi, *Some Memories About Musicians* (Tehran: Obeyd Zakani, 1996), 115; in Persian. Personal communication from Iranian scholar Majid Labbaf.

135 "He poured emotion into each of the thousand notes": Attar, *The Conference of the Birds,* trans. S. C. Nott (London: Continuum, 2000 [1954]), 26.

137 "package formation in nightingales": Henrike Hultsch and 137 Todt, "Memorization and reproduction of songs in nightingales," *Journal of Comparative Physiology A,* no. 165 (1989): 202.

139 "whistle songs have a specific signal value": Marc Naguib et al., "Responses to playback of whistle songs and normal songs in male nightingales: effects of song category, whistle pitch, and distance," *Behavioral Ecology and Sociobiology* 52 (2002): 216.

139 "a song to listen to, *but not to live with*": Lord Grey of Fallodon, *The Charm of Birds* (New York: Frederick Stokes, 1927), 72, 76.

140 He really didn't like that truck: Nicholas Thompson, "The Many Perils of Ejective Anthropomorphism," *Behavior and Philosophy* 22, no. 2 (1994): 59–70.

140 "what is joy to me, to him is pain": Oscar Wilde, "The Nightingale and the Rose," *Oscar Wilde: Complete Shorter Fiction* (Oxford: Oxford University Press, 1979 [1890]), 104.

143 you can hear it today: Many fine nightingale recordings, along with excerpts from some of Beatrice Harrison's concert and the nightingale-bombers wartime duet are all on the CD *Nightingales: A Celebration,* available from the British Trust for Ornithology, at: www.bto.org/appeals/nightingale.htm.

Chapter 7

146 only raspy, buzzing, buglike sounds: Masakazu Konishi, "Effects of deafening on song development in American robins and black-headed grosbeaks," *Zeitschrift für Tierpsychologie* 22 (1965): 584–99.

146 "the motor pattern of song . . . can be maintained": Masakazu Konishi, quoted in Marler and Slabbekoorn, eds., *Nature's Music: The Science of Birdsong,* 25.

151 this specific gene . . . is expressed whenever a bird . . . sings his particular song: Claudio Mello and David Clayton, "Song-induced ZENK gene expression in the auditory pathways of the songbird brain," *Journal of Neuroscience* 15 (1995): 6919–25.

152 "ZENK is induced": Erich Jarvis, "Brains and Birdsong," in Marler and Slabbekoorn, *Nature's Music,* 241.

156 "a new theory of long-term memory": Fernando Nottebohm, "Why Are Some Neurons Replaced in the Adult Brain?" *Journal of Neuroscience* 22, no. 3 (2002): 624–28.

156 "more fun when nobody believed it": Fernando Nottebohm, quoted in Michael Specter, "Rethinking the Brain," *The New Yorker,* 23 July 2001, 53.

157 "find out how nature uses it": Nottebohm, "Why Are Some Neurons Replaced," 627.

157 "there is no equivalent distinction": Jarvis, "Brains and Birdsong," 271.

158 "more akin to music than language": Erich Jarvis, interview with author, April 2004.

159 "a program open to new . . . information": Martine Hausberger et al., "Neuronal bases of categorization in starling song," *Behavioural Brain Research* 114 (2000): 94.

160 148 sensors to the scalp: Aniruddh Patel and Evan Balaban, "Temporal patterns of human cortical activity reflect tone sequence structure," *Nature* 404 (2000): 80–84.

162 "virtuosi in . . . group safety": William Benzon, *Beethoven's Anvil* (New York: Basic Books, 2001), 182.

164 every syllable produced by the birds was compared and analyzed: Ofer Tchernichovski et al., "Studying the Song Development

Process," *Behavioural Neurobiology of Birdsong, Annals of the New York Academy of Sciences* 1016 (2004): 348–63.

Chapter 8

172 twenty million copies: Michael Remson, *Septimus Winner: Two Lives in Music* (Lanham, MD: Scarecrow Press, 2002), 69.

180 not a fanfare but an elegy: G. R. Mayfield, "The mockingbird's imitation of other birds," *Migrant* 5 (1934): 17–19. See also Jeffrey Baylis, "Avian vocal mimicry: its function and evolution," in *Acoustic Communication in Birds,* vol. 2, ed. Donald Kroodsma (New York: Academic Press, 1982), 51–83.

181 "the female quivered her wings": Peter Merritt, "Song Function and the Evolution of Song Repertoires in the Northern Mockingbird," unpublished Ph.D. diss., University of Miami, 1985, 17.

182 up to sixteen times a second: Roderick Suthers, "How birds sing and why it matters," in *Nature's Music,* ed. Peter Marler and Hans Slabbekoorn, 285.

183 "this principle of discontinuity": D'Arcy Thompson, *On Growth and Form,* 2d ed. (Cambridge: Cambridge University Press, 1942), 1094.

184 the will of each creature to survive: Alexander Skutch, *Harmony and Conflict in the Living World* (Norman: University of Oklahoma Press, 2000), 138.

185 "thisonethisonethisone": Frits Staal, *Ritual and Mantras: Rules Without Meaning* (New York: Peter Lang, 1990), 305.

185 "our highest ethical standards": Alexander Skutch, *Origins of Nature's Beauty* (Austin: University of Texas Press, 1992), 268.

185 "the fabulous sound comparison software": by Partha Mitra, at: www.ofer.sci.ccny.cuny.edu/html/sound_analysis.html; or www.talkbank.org/animal/sa.html

Chapter 9

189 "a little fool lies here": Meredith West and Andrew King, "Mozart's Starling," *American Scientist* 78 (1990): 114.

192 a musical joke: George Grove, *Beethoven and His Symphonies* (New York: Dover, 1962 [1898]), 208.

193 "They'll reunite tonight": Rebecca Rischin, *For the End of Time: The Story of the Messiaen Quartet* (Ithaca: Cornell University Press, 2004), 10–11.

195 "the true, lost face of music": Olivier Messiaen, in *Le Guide du Concert,* 2 April 1959, quoted in Robert Sherlaw Johnson, *Messiaen* (Berkeley: University of California Press, 1975), 117.

197 "one, two, or three octaves lower": Olivier Messiaen, *Music and Color: Conversations with Claude Samuel,* trans. Thomas Glasow (Portland, OR: Amadeus Press, 1994), 95.

201 "no similar freedom could be imagined": François-Bernard Mâche, "Syntagms and paradigms in zoomusicology," *Contemporary Music Review* 16, no. 3 (1997): 77.

205 "I'm ready to sing, are you?": Fredric Vencl and Branko Soucek, "Structure and Control of Duet Singing in the White-Crested Laughing Thrush," *Behaviour* 57, nos. 3–4 (1976): 221.

Chapter 10

213 "why should he have continued practicing?": L. H. Smith, *The Life of the Lyrebird* (Richmond, Victoria: W. Heinemann, 1988), 101–2.

216 "Love the details": Edward O. Wilson, *Consilience: The Unity of Knowledge* (New York: Knopf, 1998), 54.

216 "both genes and culture must be understood": Ibid., 163.

217 "We always know more than we can say": Wendell Berry, *Life Is a Miracle* (Washington, DC: Counterpoint, 2000), 45.

218 "I don't think creatures can be explained": Ibid., 113.

218 "God is the song of the universe?": Roger Payne, *Among Whales* (New York: Scribners, 1995), 167.

220 "forgetting the names of the sounds you hear": I'm paraphrasing artist Robert Irwin, who said that "seeing is forgetting the name of the things you see," as reported by Lawrence Weschler in a book of the same name (Berkeley: University of California Press, 1983). It's an artistic reach for a kind of direct perception, the philosophy of phenomenology described by Edmund Husserl.

221 "sheer joy in life": Alec Chisholm, "Lyrebird Revels," in *Birds of Paradox* (Melbourne: Landsdowne Press, 1968), 106.

226 *The Albert Lyrebird: Prince of the Rainforest:* This impressive film is

available only from O'Reilly's Rainforest Guesthouse, at: www.oreillys.com.au

227 the lyrebird a probable ancestor: Per Ericson, Les Christidis, et al., "Systematic affinities of the lyrebirds with a novel classification of the major groups of passerine birds," *Molecular Phylogenetics and Evolution* 25 (2002): 53–62.

228 fourteen million of them: Audubon Magazine State of the Bird Report, October 2004, at: www.audubon2.org/webapp/watchlist/viewSpecies.jsp?id=222

228 doing better than ever: For an even review of the health of songbird populations see Scott Weidensaul, *Living on the Wind: Across the Hemisphere with Migratory Birds* (New York: North Point Press, 1997), 345–70.

228 Species disappear daily: Edward O. Wilson, *The Future of Life* (New York: Knopf, 2002), 186.

Illustration Credits

Page XIV: Field sketch of white-crested laughing thrush made in Nepal by Fredric Venci. Used with permission.

Pages 63, 64, 67, 68: Sonograms from W. H. Thorpe, *Bird-Song: The Biology of Vocal Communication and Expression in Birds,* Cambridge: Cambridge University Press, 1961. Used with permission.

Pages 71, 75: All images from Rosemary Jellis, *Bird Sounds and Their Meaning,* Ithaca: Cornell University Press, 1977. Used with permission.

Pages 76, 77: "Dabelsteen's Comparison of Full Song and Twitter Song in the Blackbird" and blackbird stances from Torben Dabelsteen, *Solsortens Sang Som Signal*, Copenhagen: Akademisk Forlag, 1994. Used with permission of the author.

Page 85: "Peter Szöke's Notation of a Slowed-down Song of the Hermit Thrush, Compared with a Sonogram" from P. Szöke, W. H. Gunn, and M. Filip, "The Musical Microcosm of the Hermit Thrush," *Studia Musicologica Academiae Scientarum Hungaricae* 11 (1969). Used with permission of the Estate of Peter Szöke.

Pages 94, 95: Marsh warbler sonograms from Françoise Dowsett-Lemaire, "The imitative range of the song of the marsh warbler *Acropehalus palustris*," *Ibis*, 121 (1979). Used with permission of the author.

Page 101: "Eens's Sonogram of the Sequence of Phrase Types in a Starling Song" from Marcel Eens, "Understanding the complex song of the European starling," *Advances in the Study of Behaviour*, 26 (1997). Used with permission of the author.

Page 132: "Naguib's Thrush Nightingale Phrase Sonograms" from Benjamin Greißmann and Marc Naguib, "Song Sharing in Neighbouring and Non-neighbouring Thrush Nightingales (*Luscinia luscinia*) and Its Implication for Communication," *Ethology*, 108 (2002). Used with permission.

Page 154: "Comparison of Brains," from Erich Jarvis, "Brains and Birdsong," in Peter Marler and Hans Slabbekorn, *Nature's Music,* London: Academic Press, 2004. Used with permission of the author.

Index

Boldface page numbers indicate illustrations within the text.